The Solar Solution:
Shedding light on problems you didn't even know you had

Dr. Gary Loren McCallister

Copyright © 2016 Gary McCallister

All rights reserved.

ISBN:1537648403
ISBN-13:9781537648408

The Solar Solution

DEDICATION

This book is dedicated to the sunshine of my life.

The Solar Solution

The Solar Solution

CONTENTS

	Acknowledgments	i
1	That Lucky Old Sun	1
2	The Solution	8
3	The Problem	13
4	Uses of Solar	18
5	Fundamental Principles	23
6	Solar Cookers	31
7	Lunar Chillers	35
8	Cooking	38
9	Water Quality	40
10	Electricalification	48
12	Solar Refrigeration	54
11	Time for solar	57
	Appendix A Plans	64
	Appendix B Cooking Guidelines	67
	Appendix C Recipes	69
	Other books by the author	75

ACKNOWLEDGMENTS

None of the ideas for capturing and using solar energy used in this book are original inventions of mine.   I am also not writing for persons engaged in industrial endeavors.   There are many sources and organizations that teach the "how to" of solar cooking. One leading proponent of solar usage is Solar Cookers International.

The Solar Solution

THE SOLAR SOLUTION

Chapter 1 - THAT LUCKY OLD SUN

"Up in the mornin'
Out on the job
Work like the devil for my pay
But that lucky old sun got nothin to do
But roll around heaven all day.

Fuss with my woman, toil for my kids
Sweat till I'm wrinkled and gray
While that lucky old sun got nothin to do
But roll around heaven all day

Dear Lord above, can't you know I'm pining, tears all in my eyes
Send down that cloud with a silver lining, lift me to Paradise

Show me that river, take me across
Wash all my troubles away
Like that lucky old sun, give me nothin' to do
But roll around heaven all day

Send down that cloud with a silver lining, lift me to Paradise
Show me that river, take me across
Wash all my troubles away
Like that lucky old sun, give me nothin' to do
But roll around heaven all day"

It's sure hot! I haven't seen any evidence that the sun is cooling down although I am told that it is only a matter of time before it burns out. How long that will take depends on how fast it burns. How long it burns depends on what it burns, and how it burns.

I got to wondering about that kind of thing. I suppose that says something about me. My sons get to thinking about their fantasy football leagues. I guess that's because these leagues all play under artificial light.

In spite of the fact that all plants depend on the sun, and all animals depend on plants, the sun doesn't receive a lot of attention by most of us. It comes up every morning like clockwork. I guess, now that I think about it, clocks revolve like "sun work." Either way, when something is so consistent, it is usually taken for granted.

I suppose that's why one should get up before the sun and go outside, at least occasionally. Maybe it isn't coming up anymore, and we're all living under artificial light! Hard to know since we got the satellite dish.

It is the sun that gives us the seasons. I suppose that might have been one of the first things early man had to complain about. The seasons are still one of my favorite topics to complain about.

DISCLAIMER

There is an ongoing debate concerning the production of solar energy and its uses. While numerous individuals advocate greater use of solar energy, others question this approach. In the interest of full disclosure, I will occasionally insert side bars with the opposing view.

The mountain-west is a far sunnier place than most other parts of the country. Thus, it would seem sensible to build solar collectors here, instead of where the sun never shines; like in New York City. Because there are hazards involved with solar energy that are not generally recognized, I think we should consider carefully the Federal governments blatant attempt to steal our sunshine by covering large areas of the west with solar collectors.

Oh, and the sun determines the time of day. There are very few people willing to give me the time of day, I can tell you, so I appreciate that a lot. I'll be discussing solar time a little later in this book.

I am told that the sun burns Hydrogen gas at its core and turns the Hydrogen into another gas, Helium. This process of energy creation is called nuclear fusion and produces heat and light. Someone has calculated that the sun burns about 700 billion tons of Hydrogen a second. Even though that number is less than our national debt, I still found it alarming. But I was only alarmed until I wrote out the number for the mass of the sun, which is 1989 with thirty zeros behind it! That is one million nine thousand, eight hundred and nine billion, billion, billion Kg.

Heck, if that number were our GNP in dollars, we could borrow even more money in the next couple of years. By comparison, several hundred billion tons of Hydrogen just isn't all that much. I suppose, looked at that way, the Social Security and Medicare shortfalls aren't all that much either.

So while it's nice to know that our sun won't be going out soon, it also means that things probably aren't going to cool off a lot in the future. In fact, before the sun burns out, it is supposed to get even hotter down here. I'm told that when the hydrogen is finally all burned up the sun will begin to burn the Helium. I'm not sure why the Helium doesn't burn now, already.

Anyway, when the Helium burns the increased heat will make the sun even hotter, and the sun will then expand in size to encompass the Earth, and maybe even Mars. Some scientists estimate that we only have about five billion years left. They also estimate that in just a billion years, before the sun's expansion, the sun will get so hot it burns off all the water on earth. That would put an end to Lake Powell and all the backyard pools. By then we won't be suffering so much from the heat as much as the humidity.

> "To the best of our knowledge, our Sun is the only star proven to grow vegetables." --Philip Scherrer

Apparently, the sun has been burning for quite a while already. I am sure it has been burning since 1945 based on my own personal recollections. But others, who must be terribly old, tell me that the sun

---

SOLAR ENERGY IN THE HOME

Did you know that we are exposed to a certain amount of background solar radiation from the sun? One thousand watts per square meter is about the average, and that exposure far exceeds the guidelines set for the new fluorescent light bulbs. But we can stay indoors away from direct effects of that radiation. Solar collectors, however, would concentrate this energy that would otherwise fall on the empty, barren wasteland of the west, and use it to irradiate us in our very homes.

There are no long-term studies to determine if electricity derived from solar radiation is safe for human consumption. Considering the known risks already inherent in the exposure to solar energy, and the dangerous mercury vapors in fluorescent bulbs, this gap in our knowledge seems almost criminal.

has already been burning for several billion years. See, at first everything was "without form and void," and it was dark. There was just this big cloud of, mostly, hydrogen that filled space everywhere. That was about five billion years ago, give or take a little. Can you believe it? No one wrote down the date!

Times were tougher then. Everything was uphill, and it was so cold that our tongues would stick to metal poles if you licked them. It was nine miles to school, and we had to milk the cows before we could even start . . . No, wait, that's a different story.

The five-billion-years-ago story is about when a bunch of the hydrogen atoms got very close together. Then gravity began to work and pull them even closer. The cluster of atoms got even more dense as the atoms are stacked up. The atoms began to exert tremendous pressure on the other atoms at the bottom of the pile. I'm not sure what the bottom was since everything was without form and void. But apparently, you still wouldn't want to be at the bottom of a lot of Hydrogen atoms.

> ### SOLAR FUSION
> The sun, of course, burns by nuclear fusion. Look what happened to Japan with its reliance on nuclear reactors. Further, the fuel that runs the sun is hydrogen, an extremely explosive gas. Some would argue that the sun is 93,000,000 miles away. But it only takes about 8.3 minutes for sunlight to travel from the sun to the earth. I figure that, with any kind of accident, like a sun quake and a resulting tsunami, the earth is just ten minutes away from being fried. By failing to acknowledge these dangers, the solar industry makes reasonable people wonder what else they might be hiding.

The pressure caused the cloud to get hotter and hotter until the hydrogen atoms blew up and flew apart. Electrons and protons were scattered everywhere. It was a mess kind of like the time I overheated an egg in the microwave. That thing exploded, and we had egg yolk everywhere. But then the heat from the hydrogen cloud began to cause the elements to glow just as if God had said, "Let there be light" or something.

The flying protons made a mess! Wherever they slammed into another proton, the two would stick to one another forming atoms of helium. This fusion produced even more energy in the form of light and heat. But heat causes things to expand (see note above on exploding eggs in microwaves) while at the same time gravity was pulling things together. Eventually, the expansion and contraction of atoms reached equilibrium, and the thing we call the "sun" was formed.

It gets a little fuzzy there in the middle between stacked Hydrogen atoms and a full-blown sun (pun intended). Physicists can't agree about the details, so most of us just call the whole thing a miracle. I think there is a formula for miracles someplace too, but I'm not up on my physics.

I'm told things are still fusing out there in the middle of the sun. They also tell me it is 27 million degrees F at the suns core, and the gravity has compressed the gas to about 100 times the density of most metals. So the protons that are kicked loose can't get out. Trapped, they are forced to collide with other protons, continuing the fusion. This creates even greater energy within the reaction.

But the energy has a hard time getting out of the dense core for some reason. I am told it can take thirty thousand years for the energy at the core to reach the sun's surface. That's slow! It's a little like me waking up in the morning. . .

If it really takes that long for energy to get out of the sun, the light we see from the sun is thousands of years old. Well, thirty thousand years plus eight and one-half minutes. Eight and a half minutes is how long it takes the sun's light to travel the 93,000,000 miles from the sun to earth. That knowledge is a little discouraging when I am working in the garden. I know, as hot as it is at that moment, it's going to be hotter in just a little over eight more minutes!

> **EXTENDED EXPOSURE**
> It is already known that people living close to the equator are at high risk for developing cataracts from exposure to solar radiation. Looking directly at the sun for even a few seconds can be dangerous. Yet we are to believe that watching television or computer screens powered by solar energy will be benign. Are we to believe that television, powered by solar energy, can actually be good for us? As far as I know, medical studies addressing this issue have not been planned.

Presently the sun is called a Main Sequence Star. But in about 5 billion years from now it will turn into a Red Giant. I wonder if that is a major milestone for stars, sort of like getting one's driver's license. Red Giants are said to last another few billion years, although it's not clear to me how we know that, or why you should care. Oh, and Red Giants are created when a lot of the hydrogen has turned into helium, and the helium starts to burn.

Finally, the sun will become a "White Dwarf" for about ten billion years. I think that's kind of what is happening to me as I age. My hair turns white, and I shrink. However, I wouldn't plan my vacation based on these predictions for outer space. Prophecy is notoriously difficult, unless you're a Prophet, I suppose. Physicists are especially poor at it. Anyway, white dwarfs won't be particularly good for passive solar heat anyway.

FIRE PREVENTION

When considering solar there are other things to think about. Every child knows that when naturally-safe levels of solar energy are concentrated by a magnifying glass, a fire can be started. So, who would actually feel safe with a solar collector on his roof? And who can afford the fire insurance? We would have to adopt a Universal Fire Plan, requiring fire insurance for all citizens, so that those fool-hardy people with collectors could afford their higher rates.

The truth is, a lot of facts about the sun are kind of squishy. For example, unlike rocky planets, the sun doesn't have a definite boundary. That makes it a little difficult to measure its actual diameter. I think that is the kind way of saying the measurements are pretty subjective. Usually, scientists measure the distance from the core to the edge of the photosphere to calculate the diameter.

The photosphere is the area where the gasses are too cool to radiate a lot of light. So, what the heck does that mean? Go outside this evening and determine where the light from your porch light ends. See?

The sun appears to be losing brightness at the rate of about 0.02% per year. How does that happen if it is always getting hotter? I hardly notice, except that my Seasonal Affective Disorder does seem to be coming earlier and staying longer. If we assume twelve hours of sunlight in a day, which is generous where I live, then the sun is losing brightness at the rate of 0.24 seconds a year. That may not sound like much, but if you're as old as I am it adds up. No, I won't say exactly how much, but it's noticeable.

Chapter 2 – THE SOLUTION

*"Little darling
It's been a long, cold, lonely winter
Little darling
It feels like years since it's been here*

*Chorus:
Here comes the sun,
Here comes the sun, and I say
It's alright*

*Little darling
The smiles returning to the faces
Little darling
It seems like years since it's been here*

*Little darling
I see the ice is slowly melting
Little darling
It seems like years since it's been clear."*
                - *Lennon and McCartney* -

I am always glad to see the sun, whether at the end of winter or the dawning of a new day. The sun has been greeted as a harbinger of new life since ancient man. One doesn't need a modern understanding of the heliocentric theory, or modern Laws of Thermodynamics and ecology, to recognize the sun as the giver of life to the earth.

> Finally, it is clear that solar energy is nothing more than a short-term fix to a long-term problem. Scientists, themselves, estimate that the sun will burn out in about a billion years. America, the largest consumer of energy, will be bankrupt long before then and will not be able to afford any kind of energy. Shouldn't we put our brains and resources to work seeking a safer more permanent solution to our energy needs.

A new day may dawn when we can harness solar energy. I'm not talking about photovoltaic cells, high-tech solar collectors on the space station, government initiatives for economic stimulus, or baseball caps on my bald head. I'm talking about cooking our food, creating safe drinking water, and keeping cooked food cool. Yes, solar is dawning, although that isn't anything too new.

We know that, as early as 1767 (ten years before the Declaration of Independence was signed) a Swiss naturalist, Horace de Saussure, created what he called a five-layered, hot box that reached temperatures of 189.5° F. If this happened in the Swiss Alps, it truly is impressive.

> Oooooh! Samuel Johnson is said to have said, "The pun is the lowest form of humor." Musician Oscar Levant astutely added, "If you didn't think of it first."

Another early publication (1830) tells about experiments performed by Sir John Herschel, a noted astronomer, who made a hot box while on an expedition to the Cape of Good Hope in South Africa. (I suspect sailing the seas occasionally left one with long, uninterrupted hours of thought, experimentation, and fooling around. These attributes seem to be generally missing from our busy lives today.)

He reported later: "As these temperatures [up to 240ºF] far surpass that of boiling water, some amusing experiments were made by exposing eggs, meat, etc. [to the heat inside the box], all of which, after a moderate length of exposure, were found perfectly cooked."

Interestingly, fossil fuels have been the enemy of solar research since the industrial revolution. In 1860, a Frenchman named Auguste Mouchout invented a small steam engine that ran off solar energy. He was even able to connect the engine to a refrigeration device to make ice. He was funded by the King and was awarded a medal for his work. But then, the French arranged to buy coal from England for a cheaper deal, and his funding was cut. Sound familiar?

There are many human experiences that are enhanced by scientific knowledge. But knowing that the earth's axial tilt is 23.26° does not make the winter more beautiful or Christmas sweeter. The fears and uneasy hearts that accompany dim days and dark nights are seldom mitigated by knowledge of why it is dark and knowing that it will pass. The joys of warmth and firelight are seldom enhanced by facts and material circumstances.

No, there are realms of our existence in which science has very little to offer. When disaster strikes and hopes and hearts are crushed, science has little to say. Our knowledge of science may help those who remain to live and rebuild a life, but it has little in the way of comfort. Too, when hearts are full, science does not increase the sweet emotions. Science serves mankind in a material way but often leaves us wanting more. Perhaps that is why so many scientists dabble in the arts. Grief is compounded by physical suffering and science has greatly served mankind's physical wellbeing.

It is amazing how living things tenaciously hang onto life during cold and dark times. Our experience teaches us that not everything is dead, but that life waits to begin again. If life is sacred, then the DNA code is sacred. And there is no more purified and refined form of DNA in nature than that within a seed. Seeds are the basis of life, and they have become the predominant metaphor for creativity, inspiration, faith, renewal and resurrection. In the dead of winter, seeds are the dead of winter.

This isn't the first book to be written about solar energy, although I suspect it is the best. (You'll want to be sure to read completely through the last chapter find out that Al Gore didn't invent solar energy.) The first book was written by William Adams in about 1876. It was called A Substitute for Fossil Fuel in Tropical Countries.

And lest you think solar cars are something new, you might like to know that The Solar Motor Company was formed around the turn of the last century. They produced two cars, both of which were destroyed by storms. The storm also destroyed the company. By the way, those first solar cars they built sold for $2,160.00. I am not sure if that included tax.

> ### DOG DAYS OF SUMMER
> Why do dogs get blamed for the hottest month of the year? I know their tongues hang out, and they look hot. I do the same. Dogs' tongues hang out because they can't sweat. They have very few sweat glands in their skin, and most of those are in their footpads. Did you know dogs have sweaty feet? You probably already know that dogs cool themselves by panting. By breathing rapidly, they use the moist surface of their lungs to evaporate moisture and cool themselves.
>
> Water has what is called a high "specific heat". This means that it takes a lot of energy to raise the temperature of water even one degree. So when water evaporates from the wet surface of their tongues it gives up that same amount of energy as it took to warm the water to body temperature. This eliminates the heat. Dogs also produce thin, watery saliva that evaporates easily on their tongues and this also helps cool them.
>
> However, that is not why the days of August are sometimes called the "dog days" of summer. The ancient Romans noticed that during the hottest part of the year the brightest star rose in the sky about the same time as the sun. This star, named Sirius, is the prominent star in the constellation Canis Major, and so the star is also called the Dog Star. (It isn't by chance that an important character in the Harry Potter series is named Sirius and is an Animagus as a large black dog.) The Romans thought that it was the combination of light from the sun and Dog Star that caused the heat, and so they called the hottest days of summer the "dog days" of summer. We've been calling it that for at least two thousand years now.

The Luz Company in Los Angeles, between 1980 and 1991, produced 95% of the solar-based electricity in the world. As usual, the companies demise was more related to politics than business or common sense. You can read more about that somewhere. I'm not sure where.

Other notable scientists have been fascinated by solar energy. Samuel Pierpont Langley, an American astrophysicist who later became head of the Smithsonian Institution, played with solar power. But heck, so did I when I was just a tike. I used to burn ants with my Grandad's hand lens. So why does he get press, and I get ignored?

One of the most famous American scientists, Thomas Edison, said this. "I'd put my money on the sun and solar energy. What a source of power! I hope we don't have to wait 'til oil and coal run out before we tackle that."

We're getting closer!

Chapter 3 - THE PROBLEM

*"Sunshine on my shoulders
makes me happy
Sunshine in my eyes can make
me cry
Sunshine on the water looks so
lovely
Sunshine almost always makes
me high"*

-John Denver-

OK, so, if solar energy is the solution, what is the problem?

The problem may be bigger than you think.

Pollution

The problem starts with the fact that one half, three billion and counting, of the world's population burns wood or dried dung to cook their food. Simply eating food cooked over dried dung qualifies as a problem for me. Although that is probably better than eating food cooked over fresh dung.

Lung disease is a female problem in much of the world because of cooking over open fires. The men mostly stay outside, producing dung I guess. I wonder if Eve had lung problems later in life.

I'm told that a burning cigarette creates more than four thousand distinct chemicals. I don't know how they determine that since chemicals are notoriously difficult to get one's hands on. Anyway, many of these chemicals are carcinogens, and nearly all of them can cause lung damage in some form. (This must be

---

THE DEAD OF WINTER

The leaves have fallen from the trees
The trees don't need them anymore
Colder now it's plain to see
The ground needs a blanket more
The dead of winter is at the door
The seeds of harvest have all been stored
Each awaits something more
To live again just like before

Flowers bow weary heads down
And give their children to the earth
The final petals come unbound
Giving birth to all they're worth
The dead of winter are at the door
The seeds of harvest have all been stored
Earth and stone becomes the door
Through which we pass to life restored

Wild seeds fall upon the ground
And await the coming Son
Seeds are laid in funeral mounds
To await resurrection
The dead of winter are at the door
The seeds of harvest have all been Stored
This is what the Son is for
Restoring life, the earth's Savior

true because I found it on the internet.)

It would seem that wood and dung smoke would be no different, and in many cases, may be worse. I wonder if anyone has identified all the chemicals in dung. Judging from my experiences with fresh dung, those chemicals must surely be unique. Just the amount of particulate matter in the air of the home would be a problem when cooking with an open fire. I suppose wood smoke is "cool" when roasting marsh mellows. Come to think of it, I even have a problem with smoke then, since it invariably blows in my face. Smoke follows beauty.

Deforestation

As with all things human, it doesn't end there. The wood for these fires, when it is available, often comes from local forests. The practice of using wood to cook with leads to deforestation. That's if there is a forest to deforest. If the forest has already been deforested, maybe that's when the dung comes into play. There is an upside, I suppose. The practice of burning dung might lead to "dedungification," and that would seem to be a desirable result.

Perhaps, if dung were properly utilized, it could provide a major source of the world's energy. Like a lot of things, there probably isn't a shortage of dung as much as a failure to deliver the dung where it is most needed. Perhaps a government initiative to utilize Congress in a more efficient manner could improve America's standing in the world. A new stimulus plan.

However, back to deforestation, the practice of burning wood accounts for 16 million hectares of forest destroyed annually. Also, the fires create carbon dioxide, and the trees are a major method earth has for soaking up carbon dioxide. Fires release more carbon dioxide at the same time they destroy the major source of its utilization!

> Implicit in the Christian story for the last two thousand years is the birth, life, death, and subsequent resurrection of Jesus Christ. But this is the same story celebrated by ancient traditions surrounding the winter solstice for unknown centuries. Both stories are entwined with the ideas of death and resurrection, of beginning again, the cycle of the seasons, dark giving way to light, repentance, growth, and change. Christ's story reminds us that there is light in the darkest hour. But it also speaks to us of darkness and danger even as we rest secure in the cradle. So, whether we are Christian, Pagan, Jew, or completely irreligious, the winter season can speak to our hearts in a way that science fails to do.

I'd be more interested in solving climate change issues if a proposal was made to plant trees for mopping up the excess carbon dioxide instead of proposals to limit the production of carbon dioxide by industrial methods. Half the world's people could cook with the solution.

Carbon Dioxide Emissions
Regardless of what one thinks about climate change, carbon dioxide is seldom good for humans. By the same token, carbon dioxide is great for plants, which in turn, are good for humans. Well, except broccoli. We don't want to get rid of all the carbon dioxide, or we will run out of V8 juice.

I predict that climate will change. I don't suppose there is anything easier to predict than that. People run into trouble when they start saying in which direction it's going to change. Although, come to think of it, I am not sure if climate has a direction. Don't worry too much about that. We'll let The English teachers tackle that one. That will give them some way of becoming relevant.

The contribution of these open fires to the amount of carbon dioxide released into the atmosphere is difficult to quantify. There are few people or agencies even attempting to quantify this. Conceivably, however, these open fires contribute as much, or more, carbon dioxide into the atmosphere than all the industry of the so-called industrial nations. While industrial nations curtail industry, the poor nations contribute at least equally to the carbon dioxide load through cooking over open fires.

> Most scientists now say our sun is a star. But it still knows how to change back into a sun in the daytime.

I've read where they have determined the amount of methane produced by the cattle of the world. Methane is a gas that is twenty-three times more powerful than carbon dioxide in trapping heat in the atmosphere. Fourteen percent of the earth's greenhouse gasses apparently come from cattle in the form of methane.

Methane can, of course, be burned. I guess this makes methane a kind of "pre-dung". Most of the methane actually comes from belching rather than flatulence. Can one safely smoke around cows? I guess belching is a kind of "pre-pre-dung."

But why can they quantify cow methane and they can't determine how much carbon dioxide comes from wood, or dung, cooking fires? Also, is carbon dioxide released when pre-dung is burned? As usual, more research is needed.

Erosion

When I was young, I could walk in the rain and not get wet. I could work in the sun and never sunburn my head. I could flip my long hair in a manner that I thought was dramatic and rebellious, but which my wife tells me was pathetically cute. Then erosion happened. Erosion is a terrible thing.

Without my full head of hair, even a few rain drops run down my face, into my eyes, and down my neck. I'm soon soaked. I can't do a full days work out in the sun unless I wear a hat. The hat, of course, makes work too hot, so I still can't do a full day's work out in the sun. Wait! Maybe erosion has some good points. But still, now when I flip my head the only thing that flips is my neck. It also makes a popping sound.

Imagine what it must be like for the soil when it rains. If there is a full "head of trees," so to speak, the water is interrupted and slowed. When it finally reaches the soil, it does so slowly and is gently soaked up. If there are no trees, it simply runs off, like rain runs off my hairless head. When it does that, it carries soil particles, humus, and nutrients with it washing off the topsoil that has the most nutrition for growing more vegetation.

There is even more to it than that. Forest soils are often very rich in organic matter, or humus. Humus gives the soil its capacity to absorb huge volumes of water, like a sponge, thereby preventing what is known as "overland flow." Overland flow, of course, is what carries away topsoil and nutrients.

I don't think my head ever had a lot of humus. This humus also tends to hold onto many of the nutrients such as nitrogen and phosphorous. The humus keeps nutrients out of the waterways where they promote the growth of algae and deplete oxygen levels. Perhaps that is the source of my confusion. Rain running off my head is leading to depleted oxygen levels in my brain.

Now, imagine the similarities between a bald head under an intense sun and a hillside under an intense sun. They are both hotter than if something covers them. It is impractical to place baseball caps on hills. Forests keep streams cool by reflecting or absorbing much of the sun's energy, just like my hair does for my head. Cool streams can hold higher concentrations of dissolved oxygen, and this is critical for many species of fish, aquatic insects, and my brain.

Forests also contribute just the right measure of coarse, dead wood, and leaves which minimizes stream bank erosion and enhances aquatic habitats. That is exactly why I kept my hair long as a youth! I was trying to minimize erosion.

Clean Water

The loss of forests and trees is almost always accompanied by a decline in water quality. Erosion leads to sediment load. Also, the water runoff tends to wash more dung into the water. It transmits disease at the same time it wastes dung that could have been used for cooking.

Nearly 1.2 billion people, a fifth of the world's population, do not have access to clean drinking water. Over 1 million children die yearly because of diseases transmitted by unclean drinking water.

FOLLOWING WORLD WAR II, THE FIRST OBSERVERS INTO THE WAR-TORN COUNTRIES OF EUROPE NOTICED TWO DISTINCT THINGS. THERE WERE NO TREES, AND THERE WERE NO PETS. MOSCOW WAS ALL ASPHALT AND CONCRETE. NOTHING WOODEN HAD BEEN LEFT UNBURNED. THERE WERE NO DOGS, CATS, OR EVEN BIRDS IN CAGES. ALL HAD BEEN CONSUMED

So, what's the problem to the solution?

Consider, if fuel became unavailable to a large portion of the earth's population that presently has abundant, clean energy, what would the result be? What would happen to your community if energy supplies were disrupted? How long would it be before the trees disappear, as well as outbuildings, fences and dog houses? When the water supply is fouled, disease and smoke related respiratory distress follows.

The problem is that long-term disruptions and disasters may cause problems for which we are not prepared and for which there may be no solution.

> Some people can tell what time it is by looking at the sun. But I have never been able to make out the numbers.

Chapter 4 – USES OF SOLAR ENERGY

I can see clearly now the rain is gone
I can see all obstacles in my way
Gone are the dark clouds that had me down
It's gonna be a bright bright, bright sunshiney day
                       -Jimmy Cliff-

Probably the best use of solar energy is for creating warmth when it is cold. Oh, I know what you're thinking. Brilliant! Right? Okay, so it's obvious. But the warmth is especially gratifying on a sunny winter day when the sun on your face feels so good . . . just before your frozen toes fall off. Of course, it's also the solar energy that cracks the vinyl on your dashboard, causes the steering wheel to burn your hands and blisters your bare feet when you walk out to get the mail. I'm telling you, solar energy is hot!

So how hot is it?

What's a petawatt?
A what? A petawatt? What's that? It's ten-with-fifteen-zero's-behind-it watts. What's a watt? A watt is the conversion of energy at the rate of one Joule per second. A what? A Joule. What's a Joule? A Joule is the energy expended applying one Newton over the distance of one meter. What's a . . . oh, never mind. A watt is a standard measure of electric power. Can we leave it at that? Your light bulb in the bathroom uses about sixty of them.

The only reason I brought it up is that the earth receives 174 petawatts of incoming solar radiation each day. Of course, about 30% of that is reflected into space by the upper atmosphere. But the other seventy percent warms the earth's surface, rocks, oceans and air keeping the earth's surface at about fourteen degrees centigrade. This temperature causes all things to expand as molecules move farther away from each other. Then the warmed air and water rise upward because the molecules are farther apart and lighter.

So here are three fundamentals of solar energy: solar energy generates heat, hot things expand, and expanded things weigh less and rise upwards. That's almost all you need to know. But not quite.

When the water warms, some of the molecules become a gas or vapor that is mixed in with the air. So, when the air rises, the water vapor rises with it. Then as the air rises higher, it causes convection currents where cold and warm air meet, and the molecules begin to move away from each other. That process is called "wind". Also, when warm moist air meets cooler air that hasn't been close to any hot rocks or anything, the water cools and the molecules get close together again and become water. Water is too heavy to stay in the sky, so it falls as rain. The convection currents move the water around, so the rain falls somewhere other than where it evaporated.

When water cools, it gives off heat. That causes more air to move around, and this movement creates winds, tornadoes, dust devils, cyclones, and even gentle breezes that cool the brow. The winds generated by the heating and cooling of air can also turn wind turbines to create electricity. So, wind energy is also solar energy.

It doesn't end there. Green plants use the process of photosynthesis to turn water and carbon dioxide into cellulose, sugars, and a variety of other chemical products. This mass of chemicals, taken together, is sometimes called biomass; but I usually just call it wood. So, when people cook over open fires, the cooking energy comes from solar energy, and the pollutants that cause lung disease come from the sun.

The total energy absorbed by the earth's atmosphere is astronomical! (Of course, it would be, wouldn't it?) The earth actually receives more energy in one hour than was used in the year 2005. In fact, the amount of solar energy used by the earth in one year is about twice the amount of energy that will ever be obtained from all the earth's natural and non-renewable resources. These include: coal, oil, natural gas, and uranium combined.

Of course, solar energy isn't equally distributed on the earth's surface or over a years' time. Still, even in temperate zones, significant solar energy can be utilized. Furthermore, it doesn't take expensive equipment or extensive knowledge to benefit from the suns light and heat. The government and many businesses would have us believe that the only benefit to us is through conversion to electricity in various ways. This chapter is about the various applications of solar energy available to man.

Active and passive

At first, I was confused about active and passive solar energy. I thought "active solar" energy was in play when I got a sunburn, and passive solar energy was when I sat in front of the picture window and just got hot. But that isn't quite right. See, with active solar energy the sunlight causes something to move. It's like when solar energy is causes wind, wind turns a windmill, and the windmill makes electricity. Another example would be sunlight pushing electrons around directly so that the electrons move. Moving electrons is called electricity.

The trouble with active solar energy is that it takes special equipment, and a lot of energy is lost during the transfer from solar to make it something else, like the mechanical energy in the windmill or electricity in photovoltaic cells. It's hard to justify equipment expenses and transfer losses for heating or lighting just one home.

Now, passive? I even like the sound of that. It brings to mind hammocks, shady fishing holes, afternoon naps, and honey straight from the hive. In some ways, those are pretty good descriptions. Passive solar involves techniques for using selected materials and designs that favor the collection, storage, or loss, of thermal properties. Passive solar energy can include the use of a large array of objects and activities. Because passive solar energy reduces the need for other sources of energy, these are called demand side technologies.

This book is concerned with "passivity", so it is good for reading in a hammock with a fishing pole by your side. It is even appropriate to set this book aside occasionally for a little nap.

> Did you hear about the blonde that stayed up all night to see where the sun went?
>
> It finally dawned on her.

It's interesting how confused people can get about active solar energy. Those people who promote active solar energy say that it increases the supply of energy, so economists call it "supply side" technology. Of course, that is nonsense from a scientific point-of-view.

The amount of energy entering the system is fixed. It is the amount of energy received from the sun. That amount of energy cannot be increased. What photovoltaic cells and windmills do is change the solar form of energy into electrical energy. So, capturing solar energy doesn't supply any additional energy, it simply takes whatever energy is available from one source and turns it into another form.

However, whenever one form of energy is converted to another, a certain amount of it is lost, usually to heat. So, the conversion doesn't increase the amount of energy. It decreases it. I concede that active solar may increase the amount of electricity. However, if one calculates the loss of photosynthesis from the erection of a field of photovoltaic cells, there may actually be a loss of net energy.

Humans have already reduced the amount of solar energy available to the world through "monocultural farming." In the natural world, that is a "vacant lot" where many different plants grow in the same area. The growing season is staggered, and there is always something photosynthesizing using the energy from the sun. In monoculture farming, one crop is grown and then removed. Even if two crops are rotated, the ground is barren with no, or little, photosynthesis occurring for the part of the year between crops.

If we want to efficiently capture more of the available solar energy entering our system, we should plant a greater variety of plants at the same time in the same place. Of course, humans aren't interested in the world's energy supply. They are more worried about their money supply, and how best to turn that free solar energy into cash flow.

That's explains why I wrote this book. I am taking nothing but dead trees, manufactured from sunlight, and hot air, manufactured personally, and creating fascinating and helpful reading advertised to help save the world.

Applications

The following section details some of the ways passive solar energy can be used. By considering these, we can learn a lot about how to think "solarly," thus being able to generate more ideas about what that you might like to pursue in your life.

I was disappointed to learn that I had not discovered the principles of using solar energy. It's kind of like all the good things have already been invented by Al Gore. Solar energy can make contributions in many ways. It can be used to warm our homes for example. However, building solar buildings, lighting with solar, and making electricity from solar energy are complex subjects that requires a much broader discussion than is possible here.

This book has far grander goals. I have the audacity to write an entire book on how to use solar energy for little personal projects that can free you from the tyranny of government, the demands and manipulations of huge corporations and the banking/financial systems, and allow you to survive more easily the coming, inevitable crash back into primitive living.

Seriously, solar energy can make major contribution towards reducing the amount of open fire cooking for a family, as well as improved respiratory health and provide clean drinking water. Further, in the case of emergency or primitive living conditions, it can be used to cook food when other supplies of fuel are not available. Most importantly, it's use is fun and surprisingly easy.

The remainder of this book will cover such impractical topics as
1. The fundamental principles of solar energy
2. How to cook with solar energy.
3. How to sterilize drinking water using solar energy.
4. How to use solar energy for chilling water or food.

> I'm such a lousy cook that my cat has only three lives left.

Chapter 5 – FUNDAMENTAL PRINCIPLES

There is sunshine in my soul today,
More glorious and bright
Than glows in any earthly sky,
For Jesus is my light.
O there's sunshine, blessed sunshine,
While the peaceful, happy moments roll;
When Jesus shows His smiling face
There is sunshine in my soul

Fundamentals

Why are books entitled "The Fundamentals of Whatever" always about a thousand pages long? Do I really have to know all that before I can even get started on advanced work? If those are the fundamentals, how thick is the Book for "Advanced Whatever"? I think authors and publishers need to think about what a fundamental is.

> "Turn your face to the sun and the shadows fall behind you."
> --Maori proverb

"Fundamentals" are those things out of which everything else is constructed. It turns out that is actually, a very short list of things. After all, the entire universe is composed of just over one hundred elements, and most of those are available only in trace amounts. I guess that's why most people think they learned everything they need to know in kindergarten. Or maybe they just don't know very much.

Fundamentals vary from discipline to discipline, of course. In math, numbers are fundamental. I admit it's an infinite set. Still, what one can do with "x", can pretty much be done to anything else.

Tones are fundamental to music. Do you realize that all the western music ever written is done with the same twelve tones? We just arrange them in different ways. Phonemes and their visual representations, letters, are fundamentals of a language. All of Shakespeare and the subway graffiti is done with the same 26 letters.

It seems that if we have any two concepts that can be differentiated from each other, we can say or build anything. We first realized the strength of that concept with the invention of Morse Code, a system of coding information using only dots and dashes. Today, I type out these words using a machine that codes my words into electronic ons and offs. This is called binary language and it is the code used for all computer-generated information.

Principles of solar energy

Solar energy consists of the light and heat emitted by the sun in the form of electromagnetic radiation. Although I could go into a technical discussion about electromagnetic radiation and how it is converted into solar energy, this is not something the average person needs, or wants, to know. Well, anyway, I don't need or want to know all that. In fact, I don't.

There are a few things you probably should know to be able to make solar decisions. I was about to say "sound" decisions, but solar is quiet; and I was afraid it might confuse the reader. Well, even more than I usually do.

The feasibility of using solar energy depends on the amount of available sunlight in your geographic area. Every part of the Earth is provided with sunlight during at least one part of the year. I say, "part of the year" because the north and south polar caps are each in total darkness for a few months of the year. The amount of sunlight available is one factor to take into account when considering using solar energy. Another factor that is important is how "passive" you want to be.

Of course, there are a few other factors which need to be looked at when determining the viability of solar energy in any given location. These are as follows:

* Geographic location
* Time of day
* Season
* Local landscape
* Local weather

Because the Earth is round, the sun hits its surface at different angles, in different locations around the globe. The angle ranges from 0°, above the horizon, to 90°, directly overhead at and near the equator. An example of 0° of sunlight would be the north pole during winter.

When the sun's rays are vertical, directly overhead, the Earth's surface gets a maximum of solar energy. The more slanted the sun's rays are, the farther they must travel through Earth's atmosphere before reaching the surface, becoming more scattered and diffuse as they go along.

- Direct and Diffuse Sunlight

The more scattered and diffuse the sun rays are, the less concentrated the solar energy is. Because the Earth is round, the polar regions never get direct sunlight, and during their respective winter months, they receive no sun at all.

The Earth travels around the sun in an elliptical orbit. Because of the Earth's elliptical path, the northern hemisphere is closer to the sun during one-half of the year, and the southern hemisphere is closer to the sun during the other half of the year. When one part of the Earth is closer to the sun, it receives more concentrated solar energy. This time of year is referred to as "summer" throughout the world.

But regardless of summer or winter, the 23.5º tilt of the Earth's axis plays a larger role in determining the amount of sunlight striking Earth at any particular location. The Earth's tilt results in longer days in the northern hemisphere during one-half the year, and longer days in the southern hemisphere during the other half of the year.

Areas such as the United States and Europe receive more solar energy between May and September, not only because days are longer, but also because the sun is almost directly overhead during this season. The sun's rays are far more slanted during the shorter days of the winter months. Cities such as Denver, Colorado, receive nearly three times more solar energy in June than they do in December.

As sunlight passes through Earth's atmosphere, some of it is absorbed, scattered, and reflected. The following is a general list of materials which cause the sunlight to become diffused:

* Air Molecules
* Water vapor
* Clouds
* Dust
* Pollutants
* Closed eyelids
* Sunglasses
* Drapes
* Window blinds
* Shade trees

Not all of these are significant when considering how to use passive solar energy. However, sunlight affected by these things is referred to as diffuse solar radiation or diffuse sunlight. Sunlight that reaches the

Earth's surface without being diffused would be called direct beam solar radiation. But there isn't much of that.

The total of all diffuse and direct beam solar radiation in a given location is called total global solar radiation. I don't know what else you would call it, but scientists like to give fancy names to things like that. Atmospheric conditions can reduce direct sunlight by 10% on clear dry days and as much as 90% on thick cloudy days. Closed eyelids can almost reduce solar radiation by 100%. This can be easily achieved if you take a nap in a shady location while your food is cooking in a solar oven.

- Measuring Sunlight and Solar Energy

Scientists can measure the amount of sunlight available at any given location and at different times of the year. So can I. So can almost anyone. There is not much to measure in January, when it's cloudy, regardless of where you are. Scientists have concluded that available sunlight is about the same at places with similar latitudes and climate conditions. Amazing! I actually read that claim in an article on solar energy.

Scientists usually report the amount of heat available from solar either as kilowatt-hours per square meter or as watts per square meter. At least that is what they use when calculating the electrical output. If they are heating water or the internal space of a building, they usually use something called British thermal units per square foot.

Are you ready for the cool thing you should know about all this? **None of it!** The fundamentals of passive solar energy are much simpler. However, like all fundamentals, they can be assembled in a wide variety of ways to create new combinations. But, there are only four fundamental concepts:
- collect light (heat),
- absorb light (heat),
- retain the heat,
- eat.

Fundamental One - Collect the Light:
Cones

Without the light, there is no heat. However, it takes a surprisingly small amount of light to work. We don't usually think in terms of a "volume of light." We tend to speak of brightness. But the amount of light available in a given space would be called a volume. Please pass a quart of light sounds strange, doesn't it?

Think of it this way, if you focus light with a lens on a given point, it will form a cone. Anyone who has ever eaten an ice cream cone knows that a cone has a volume and that one can only increase the volume by increasing the height of the ice cream. The diameter is fixed by the physical diameter of the opening to the cone.

So, if you want to know exactly how much an ice cream a cone will hold you can calculate the volume of a cone with the formula: volume = 1/3(Area of Base) (Height), or 1/3 (pi x r^2) (h). The opening of a cone is called its base and is, by definition of a cone, circular. So you must calculate the area of the circular base first.

The distance around a circle is called its circumference, and the distance across the circle through its center is called its diameter. I am sure all of you recall that if we divide the circumference of a circle by its diameter, we get the same number every time called pi. You don't? For Pete's sake, we covered that in the eighth grade! Well, anyway, the number pi is approximately 3.14. Does it make you feel any better to know that I couldn't recall any of this either and had to look it up on Google?

The reason you need things like diameter and pi is to calculate the area of the base of your cone so you can select the cone that holds the most ice cream. Knowledge is power! But wait! We aren't done. To calculate the area of the base you have to multiply pi times the radius. The radius is one-half the diameter. The height part is pretty simple, though. You just measure that. Then you multiply the area times the height, and then multiply that number times one-third. Don't ask me why the one third part.

Of course, no one seems to make the pointy ice cream cones anymore, which messes everything up. But, when you focus light it still makes a cone, so you can still calculate the volume of light in your cone. For example, if you had a cone with a diameter of about 20 inches the area of the base would be 314 square inches. If your cone were forty-eight inches high, the volume would be between about 5000 to 6000 cubic inches, depending on how accurately you measure everything. (I know, scientists usually measure in millimeters, but this column is for real people.) And if inches are too difficult, twenty inches is about the length of most people's forearm from elbow to the tip of your fingers.

If you have 5000 cubic inches of light for several hours, you can probably cook food with a cone shaped solar cooker. If you make the base wider, or the height taller, you will get more light and hence more heat. But increased volume also takes longer to heat up, so it is a trade-off. The above dimensions work well.

The advantage of using a cone is that, if there is a reflective surface on it, like tinfoil or something, it tends to focus the light into the tip. This results in a higher temperature at the focal point for better cooking.

Of course, If you are more interested in ice cream, you probably shouldn't put it in this kind of cone. With ice cream, it is probably easier to change the height of the ice cream than the diameter. I'll have three scoops, please. But for my baked potatoes, please pass eighty-seven quarts of light.

Cubes

Of course, one doesn't have to cook using a cone. Most cooking in the kitchen is done in, or on, four-sided cook stoves, or cubes. One can cook in a solar cube, and the ideas are the same as with a cone. You need a big enough opening to collect enough light, and the size of the opening remains similar. Choosing the side of the cube is a tradeoff between allowing your cooking pots to fit in, and keeping the space small enough to heat efficiently. Many sizes and shapes work.

One can increase the size of the opening on a cube by using reflectors. Many commercial systems have hinged or attachable reflectors that increase the surface area for collecting the light. However, you can make your reflectors out of cardboard covered with aluminum foil, mylar, or almost any shiny material. I don't like glass and mirrors because they focus too much light and can cause eye damage or burns. The other substances reflect light more diffusely and give a more uniform heat.

Fundamental Two – Absorb the Light:

Look close. Well, you don't have to look too closely to see that most leaves are dark in color. Usually they are green, but they are often other colors too. Leaves are often green to absorb the light. We see green because all the other wavelengths are being absorbed, and only the green wavelengths are reflecting back at us. However, all that has to do with photosynthesis, and not solar cooking, cooling, and energy production. So you can skip this paragraph if you want.

However, while there is an interesting variety of color in plants, I have never seen leaves that resemble aluminum foil. That's because, after collecting the light, the light energy needs to be absorbed into something to be useful. When all the light waves are reflected back, we see white light. When all light waves are absorbed, we see black. Since we want to absorb the light after collecting it, we want everything in the system after the collectors to be black. Well, at least dark colored.

A lot of dark-colored cookware is also quite massive, such as cast-iron cookware. Heavy cookware is not particularly good for solar cooking because it absorbs so much energy heating the mass of cookware. It slows the heating of the food, which is the point in the first place. Thin-walled metal, glass, or thin-ceramic containers cook faster. When purchased these can already be a dark color, but you can use normal cookware by just painting it a dark color on the outside. Flat paints work better than enamels. In fact, dark tempura paints can even work, or you can coat the cookware with paste out of ashes and flour. That's messy, but it works.

Uniform exposure to light is sometimes a problem. It can cause food on top of the cooking container to be cooked while food on the bottom is not. Placing the cooking pot on top of a screen, or a lift, allows light to penetrate and reflect off the bottom below the pot to achieve more even cooking.

Fundamental Three – Retain the Heat:
I've always been pretty good at getting a job. I've made a modest amount of money. The problem seems to be in holding on to the money after I get it. If I made a lot of money, this would probably not be much of an issue. But with moderate income, it is easy to lose money faster than I make it.

So instead of money, let's consider heat from sunlight. It's not hard to get heat from the sun. The temperatures are much more moderate than cooking with fire. The cooking vessels are often thin-walled for collecting the heat in the first place. So it can sometimes be a problem if heat is lost faster than it's collected. With solar cooking, heat must be trapped and retained once it has been trapped and retained.

Trapping Heat: This is usually done by applying some transparent material above or around the cooking vessel. This transparent material can be glass or plastic of some kind. Polyethylene plastic tends to melt and sag when it gets hot. If it touches the container, it will probably melt a hole in it. There are a number of plastics that can withstand heat. Glass is excellent but breakable and expensive. Rigid plastics and corrugated plastics work well.

In recent years a special plastic has been developed that is used to cook turkeys or other large pieces of meat. These are usually called "cooking bags", and they are large enough to wrap around your cooking vessel. It will not melt or burn at high temperatures. In most cases, you can simply invert the bag over the cooking vessel like a tent. An added benefit is that they are reusable.

> Why does a solar chef make breakfast at noon?
> That's the only time it can cook eggs sunny side up.

Safety

The cooking vessel gets surprisingly hot when in use. There are no usual sizzling or boiling sounds that are cues to heat. But solar cookers can reach temperatures of 250° F, and occasionally hotter. Don't forget the potholders!

The only other safety issue when cooking with solar is to be aware of reflected light. Most solar cookers, such as box cookers, cones or panel cookers, reflect light in a diffuse manner, so light is seldom an issue. I suppose, if one were working near the cooker for a long period of time the reflected diffuse light could cause some eye damage. But since the cooker is usually outside and the cook is usually out of the hot sun, this is seldom a problem.

The one exception is with parabolic solar cookers. Parabolic solar cookers use large, mirror-like surfaces that focus light on a small field where the cooking vessel is set. These focused fields can get as hot as 700° F. This is far hotter than most kitchen ovens. This extremely high temperature makes cooking on these a little trickier, requiring more intense supervision. The reflected light from the cookers can cause severe burns. Then if you look directly into them, the light can damage your eyes. That's why I don't include discuss this type of solar cooking in this book. But if you start looking around you will find articles about parabolic cookers. I just think solar cooking should be safe and easy.

> What happens to a chicken left in a solar oven for too long?
> ...It gets sun burned skin.

Chapter 6 – SOLAR COOKERS

You are my sunshine, my only sunshine
You make me happy when skies are gray
You'll never know dear, how much I love you
Please don't take my sunshine away
　　　　　　　　　-traditional-

How Making Use of Solar Energy Can Keep You from Becoming Half-Baked
- When building a solar cooker, you'll be too busy to yell at your kids.
- Building solar cookers requires concentration so you can't think about the world news.
- It keeps you humble. Tools always keep you humble.
- Using a cooler makes the dog sing along. No, wait, that's the harmonica.
- The cooler can be used to signal someone when you are lost in the desert.
- It's inexpensive to make and use, so you have money left to buy ice cream.
- Cooking with solar allows you to multitask without being in a hurry.
- You cannot burn the dinner.
- Cook time requires long nap periods before dinner is ready.

Over the past two hundred plus years, numerous types of solar cookers have been created and used. Today, there are about four basic types in use.

<u>Solar Ovens or Box Cookers</u>: These are the kind that Saussure first used, and others later experimented with. They consist of a large, insulated box with a clear top to allow in the sun. Sometimes reflectors are incorporated to increase the amount of sunlight entering the box.

These are easy to build and inexpensive, characteristics which have made them popular. They provide slow, even cooking of larger quantities of food. Their disadvantages are that they do not get as hot as other ovens and, so, cook slower. They must be turned periodically throughout the day to keep them facing the sun. There are suggested instructions in the appendix.

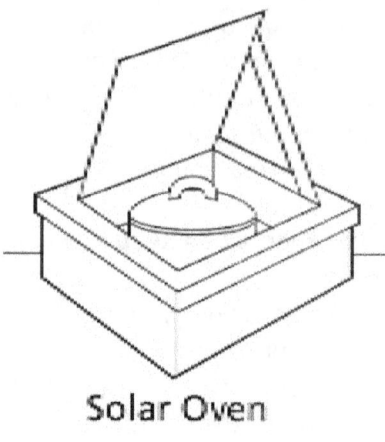

Solar Oven

Panel Cookers: These cookers utilize double pots, one clear and one dark, surrounded by a set of reflectors. The reflectors can be easily built at low cost, but the double pot can be difficult or expensive to obtain or build. Using a cooking bags for the clear outer pot makes these more flexible and affordable. The limitations of the doubled cooking vessels and the difficulty of keeping the reflectors focused on the pots make this type slightly more cumbersome to utilize. Once again, the apparatus needs to be turned frequently to face the sun. Using a panel cooker is also a little more difficult to cook larger quantities of food. There are suggested instructions for use in the Appendix.

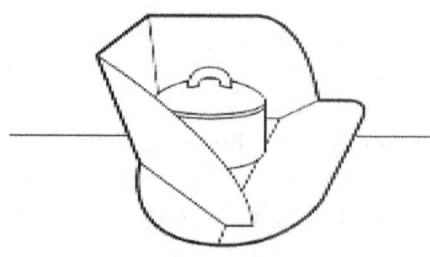

Panel Cooker

Parabolic Reflectors: These are made with a concave, reflective lens that reflects light to a specific site. The cooking pot sits on a stand at the point of the reflected light. These are more difficult to build because they must be sturdy and must be precisely focused. Because they can achieve very high temperatures they are not good for cooking. The cooking vessel tends to cook unevenly, the high temperatures achieved make it a little dangerous, and the reflected rays can burn the skin, or the eyes of a bystander. These also need to be rotated periodically to be effective. As previously stated, I have not included instructions for building this type of solar cooker.

> What happens to a goose in a solar oven when
> the sun goes behind the clouds?
> ...It gets goose bumps.

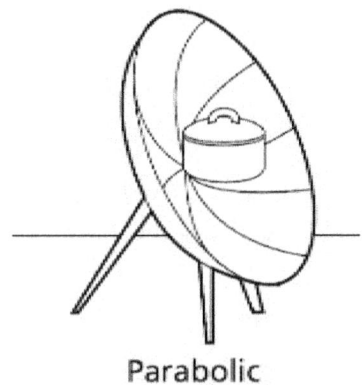

Parabolic
Solar Cooker

Funnel Cookers: These are easily and economically constructed cones that have a reflective surface. The dark-colored cooking vessel is placed in the bottom of the cone with a clear cover or a cooking bag over it to trap the heat. Because the cooker is a cone shaped, it reflects light from the sides and focuses the light onto the cooking vessel but in a more diffuse manner. If this cone has a wide enough opening, it never needs to be turned to collect the sunlight. Instructions for building such a funnel can be found in the index. The funnel cooker does need some kind of stand, but a cardboard box works, as well does a hole in the ground.

> Q: Why do people like cars
> with sunroofs?
> More leg room.

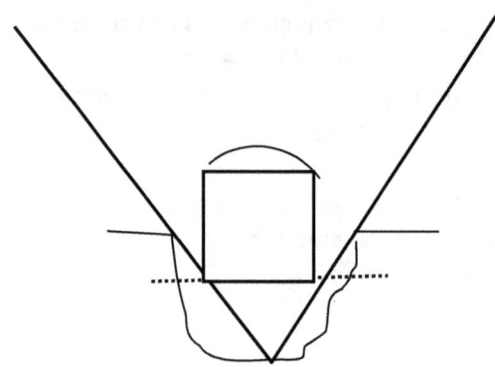

FUNNEL COOKER

What is a solar cook's favorite day to cook?
...Sun-day, of course

Chapter 7 – LUNAR CHILLERS and SOLAR COOLERS

Is this the little girl I carried?
Is this the little boy at play?
I don't remember growing older
When did they?
When did she get to be a beauty?
When did he grow to be so tall?
Wasn't it yesterday
When they were small?

Sunrise, sunset
Sunrise, sunset
Swiftly flow the days
Seedlings turn overnight to sunflowers
Blossoming even as we gaze
Sunrise, sunset
Sunrise, sunset
Swiftly fly the years
One season following another
Laden with happiness and tears

The concept of solar cooking can also work in reverse. Box and Panel cookers won't work for lunar chilling, but the funnel cooker works well. If one takes a container of water, places it into a solar funnel cooker, places the solar funnel cooker outside <u>on a clear night</u>, and points the funnel at the clear night sky, the process will work in reverse. Heat from the water will be liberated and reflect off the funnel into deep, cold, space, and the water will chill.

The chiller needs to be set in an open space without trees, hillsides or buildings blocking full access to the night sky. In the morning, the water will almost always be several degrees cooler than the ambient air temperature.

In this way, it is possible for one to pasteurize safe drinking water during the day, and chill it at night. Wrapping the chilled water in a blanket will keep it cool for many hours and can also help chill and preserve crisp, fresh fruit and vegetables. The chilled water can be used for chilling other foods during the day by placing it in an insulated box or chamber. Sometimes just wrapping food with the chilled water, in a blanket, can create a cold box.

> If a solar oven is good for cooking in the sunny day time, then what is a lunar oven good for?
> ...A night cap?
> ...A Midnight snack?
> ...Cooling your Moonpie?

LUNAR COOKER RECIPES

Moonlight Breakfast Salad

Yea, I know. Salad for breakfast? But try it, you might like it. Just before setting out in the lunar cooker, mix your favorite salad ingredients in a bowl such as lettuce, peppers, carrots, cabbage, cucumber other vegetables and spices. You could even add grapes or other fruits.

Place the bowl in the lunar chiller and set the chiller outside. Be sure it is set away from tall trees and building and is oriented towards deep space.

Do not use the lunar cooker on nights with cloud cover.

While waiting for the salad to chill, hard boil eggs and dice ham. You can also use crumbled fried bacon. (Sausage does not go well in Moonlight Breakfast Salad.)

In the morning, when the salad is crisp and fresh add the diced ham and hard-boiled eggs and serve.

Do not use your lunar chiller when the temperatures outside are below freezing. Well, unless you want frozen water. There are other selected recipes for lunar chiller entrées in the Appendix.

SOLAR COOLERS

One problem in many parts of the world, and during difficult circumstances, is to cool foods after cooking so that they remain safe to eat. Refrigeration is difficult without electricity. However, there are some methods of using solar to keep things at least cooler, if not actually cold.

Although it takes a large amount of heat to increase the temperature of water, when water evaporates it give up the same amount of heat. Thus, evaporation can be used in some instances to cool the environment.

THE SOLAR SOLUTION

Chapter 8 – COOKING

Busy old fool, unruly sun,
Why dost thou thus
Through windows, and through curtains call on us?
Must to thy motions lovers' seasons run?
 -John Donne-

Solar cookers of all types can reach temperatures on a sunny day of 130-145°C (265-300°F). On a partly cloudy day, the cooking vessel should still reach 95-105°C (200-220°F. These are the temperatures of a warm oven, or a slow cooker and are hot enough to burn you. Use pot holders. Further, if you have trapped your heat, and the sun goes behind a cloud, the food will continue to cook for at least twenty minutes.

There are more extensive guides for cooking times in Appendix B. But here are some general guidelines for judging cooking times.
- Vegetables: 1.5 hrs
- Rice/Wheat: 1.5-2 hrs
- Beans: 2-3 hrs
- Meats: 1-3 hrs
- Bread: 1-1.5 hrs

Like for almost any kind of cooking, it is hard to describe exactly how to judge these things. Trial and error is probably essential to becoming comfortable with the process of solar cooking. For example, we have learned that it is hard to cook vegetables using solar because the temperatures are so low, they cook so slowly that they often become mushy. If they are part of a soup or stew that works. If they are to be a side plate of peas, not so much.

However, like a slow cooker, we have discovered that it is almost impossible to burn food in a solar cooker. Also, because the pot is always covered to retain the heat, one needs less water than usual for recipes. It follows that since the water cannot readily escape, and the temperatures are relatively low, the food simply never burns. The ideal is to put the food on early and then let it go until you are ready to eat.

We have cooked meats of all kinds, lasagna, potatoes, stews, bread, cereals, and cakes. There are sample recipes listed in Appendix C.

One should, also, keep in mind that once food has reached a given temperature, it can be wrapped in a blanket, buried in the ground, or covered with vegetation; and it will continue to cook for at least another hour or two. Even if the day turns cloudy or rainy, dinner need not be ruined. In fact, if you can get twenty or thirty minutes of every hour with full sun, you can cook with solar.

However, the typical method of solar cooking is to prepare food early and get it out in the sun. A few passing clouds will not matter. Nor will it matter if it stays bright and hot all day. The food will not burn and will almost surely be done by time for the evening meal.

Chapter 9 – WATER QUALITY

Inflatable pool full of dad's hot air
I was three years old
Splashin' everywhere
And so began my love affair
With water

On a river bank
With all my friends
A big old rope tied to a limb
And you're a big old wuss
If you don't jump in
The water

Daytona Beach on spring break
Eighteen girls up on stage
White t-shirts about to be sprayed
With water

*Yeah when that **summer sun** starts to beatin' down*
And you don't know what to do
Just go and grab someone you wanna see in a bathing suit
And drive until the map turns blue
    - Brad Paisley -

Obviously, it isn't just any old water that will do. We need quality water. The problem is that all living things require water, even those things that aren't very good for us, like Cholera. However, that "Lucky Ole Sun" can help us with the problem of clean water as well as clean cooking.

> A minister is stopped by a state trooper for speeding. The trooper smells alcohol on his breath and sees an empty wine bottle on the floor.
> The trooper asks, "Sir, have you been drinking?"
> And the minister says, "Just water."
> The trooper says, "Then why do I smell wine?"
> And the minister looks down at the bottle and says, "Good Lord, He's done it again!"
>
> (Disclaimer - solar stills cannot do this.)

In the 1800's John Tyndall discovered that if he heated to boiling a hay infusion, organisms would grow in it again. He did not know about heat-resistant spores at the time, but he continued his studies. He eventually learned that if he reheated the soups repeatedly, he would eventually kill all the organisms in it; and a sterile broth would remain.

Louis Pasteur utilized this concept to sterilize media for growing infectious bacteria. He discovered that all the common pathogenic bacteria that were of concern could be killed by simply raising the temperature to 65° C (150° F). This process came to be called pasteurization. It has been widely applied to milk products to make them safe for distribution.

The concepts of milk pasteurization can also be applied to water. If one can raise the temperature to the boiling point, the water is rendered safe to drink. That is, the bacteria and rotaviruses - the main causes of severe diarrhea in children - are rendered inactive. After boiling for two minutes, all pathogenic microbes are inactivated. Since solar cookers routinely reach these temperatures, water can be made safe to drink by solar cooking.

Filtration

Surface water is often contaminated by dirt, organic debris and potential pathogens. Probably the first thing that always needs to be done is to filter the water. There are numerous kinds of commercial filters on the market. However, one can often make contaminated water more palatable simply by filtering it through several layers of cloth. This at least removes the big chunks.

One can make their own filter with a combination of coffee filters, charcoal and aquarium filter floss. Simply layer these three substances in multiple layer in something like a plastic pitcher. Poke holes in the bottom of the pitcher to let the water flow through and you will have a homemade filter.

Okay, so this doesn't require any solar energy, but it is still the first important step in making safe drinking water.

Direct Heat

Safe water can be accomplished in numerous ways. A dark colored container full of water can be placed within a transparent container. Or in the case of the funnel cooker or panel cooker, the cooking vessel could hold the water and be covered with a cooking bag. Any arrangement of two containers, an outer clear container and an inner container would work. An aluminum pop can in a polyethylene pop bottle would work.

The problem with solar water purification is making sure the water reaches the necessary temperatures. A low-tech device has been developed that can provide that information. It is called a W.A.P.I. (Water Pasteurization Indicator). Fred Barrett, with the United States Department of Agriculture, came up with the idea for using vegetable wax, with a melting point near 70°C (158°F), as an indicator. He built several models using wax inside a plastic cylinder and successfully used them to verify pasteurization conditions of contaminated water. In 1992, Dale Andreatta -- a Ph.D. candidate in mechanical engineering at the University of California, Berkeley created

---

A physicist, biologist and a chemist were going to the ocean for the first time. The physicist saw the ocean and was fascinated by the waves. He said he wanted to do some research on the fluid dynamics of the waves and walked into the ocean. Obviously, he drowned and never returned.

The biologist said he wanted to do research on the flora and fauna found on the ocean floor and walked into the ocean. He too, never returned.

The chemist waited a long time and then afterwards, wrote the observation, "The physicist and the biologist are soluble in ocean water".

the WAPI in its current form.

The WAPI is a clear-plastic tube partially filled with a soybean wax that melts at 70°C (158°F). With the solid wax at the top end of the tube, the WAPI is placed in the bottom of a black container of water that is solar heated. If the wax melts and falls to the bottom of the tube, it ensures that water pasteurization conditions have been reached. The WAPI has a stainless-steel washer around it to hold it at the bottom of the container, the coolest location when solar heating water.

Stringing a fish line through holes in the pinched ends of the tubing, as indicated in the illustration, the WAPI can be reused by simply reversing the direction it hangs in the water. The string also allows for removing the WAPI without contamination the water with hands.

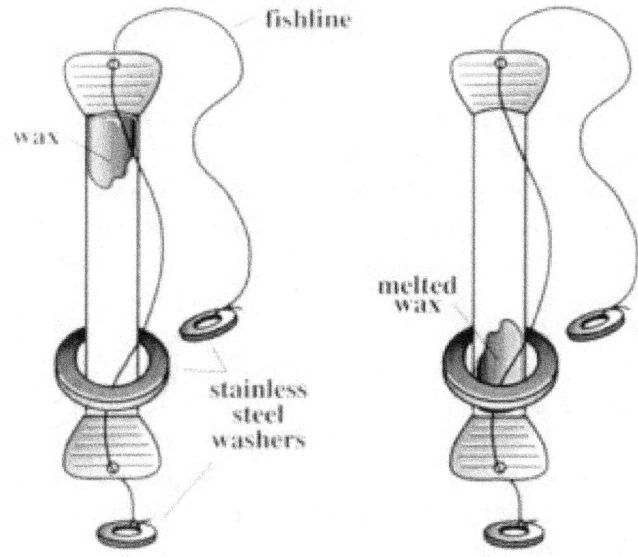

WAPI's can be constructed from common materials by simply pinching off the end of a piece of polyethylene tubing using heat. Add appropriate wax, seal the opposite end, and poke a hole through the pinched-off ends for the fishing line and then add the appropriate washers. The necessary soybean wax is probably not available locally and would have to be ordered.

Be sure the tubing is completely sealed. If water gets inside, it will alter the melting point of the wax and make the readings unreliable. WAPIs can also be purchased from Solar Cookers International:
http://solarcookers.org/

http://65.108.108.197/catalog/waterpasteurization-c-23.html

Solar energy can also be used to create stills. There are numerous designs available on the internet. As a long-term project for larger groups, these might be worthwhile building. But for short term survival and safety, pasteurization probably is the simpler method.

> Water, water, everywhere,
> And all the boards did shrink;
> Water, water, everywhere,
> Nor any drop to drink.
>
> Rime of the Ancient Mariner,
> Samuel Taylor Coleridge

Solar Stills

Safe water can also be obtained by distilling contaminated water. Distilling is the process of heating a liquid until it becomes vapor, then cooling the vapor back into a liquid on a separate clear surface and container. It has been popular for making alcoholic beverages for centuries. In fact, during much of the history of the world the water was so unsafe that drinking alcoholic beverages was the only really safe drink available. Thankfully, we can now make solar stills and provide clean drinking water. Well, some people would be thankful I guess.

Because water has such a high specific heat, which means that it takes a lot of energy to raise its temperature, distilling requires a lot of energy. This makes it difficult to produce large amounts of water by distillation. However, distilled water is the safest form of water from an infectious perspective.

Solar energy can be sufficient for distillation using some very simple mechanisms. The problem will often be producing a large enough quantity to satisfy needs.

The simplest solar still can be as little as a clear two-liter bottle and an empty aluminum pop can. Simply cut the upper one third of the plastic bottle off. Place the aluminum can full of water inside the bottom half. Cut one or two vertical cuts about an inch long in the top part of the plastic bottle and replace it over the can, but wedging it down so that it fits inside the bottom of the plastic bottle.

Place this still in direct sun. As the dark colored aluminum can heats the water, the water will vaporize. When the vapor strikes the cooler outer bottle, it will condense into droplets and run down the sides and collect in the bottom of the bottle. You can later remove the top and the can and drink the distilled water in the bottom. This process can be enhanced by placing the bottle in an area with reflectors to concentrate the sun.

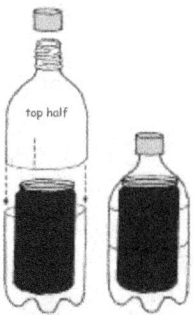

Obviously, a solar oven as discussed previously could also be used to process larger volumes if there is a clean method and container to collect the concentrate.

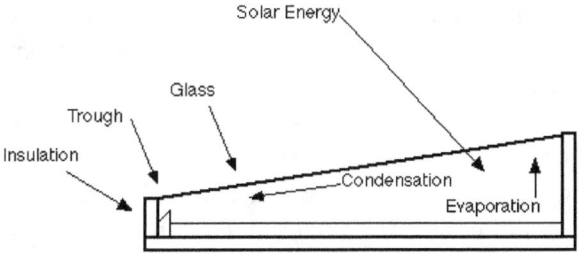

Desalination

Salt is important stuff. Salt is essential for human life, and saltiness is one of the basic human tastes. It is one of the oldest and most ubiquitous food seasonings recorded. Before refrigeration, salting was the most significant method of preserving food. Salt extraction and mining have been recorded from as long ago as 8000 years in both Romania and China. It has been used for money and trade for centuries. The significance of salt has even flavored our language.

The phrase "Salt of the earth has been used in English to refer to someone who is ordinary and unsophisticated but of a decent sort. It was used as a metaphor in the Sermon on the Mount by Jesus Christ to represent something of value such as light. If salt has lost its savor, it is not good for anything. The phrase "salty dog can refer to a salty drink made from Vodka, an experienced sailor, or one with a large sexual appetite. Unfortunately, usage is never exactly clear which is which in this latter phrase.

However, there is one place where almost all agree that no one wants any salt is in our drinking water. This is because the human kidneys can only make urine that is less salty than salt water. Therefore, to get rid of all the excess salt taken in by drinking seawater, you must urinate more water than you drank. Eventually, you die of dehydration even as you become thirstier.

There is a lot of water on earth. It is estimated by someone, not me, that there are 333 cubic miles of surface area of water on earth. However, it is not evenly distributed. Saltwater accounts for 97.5% of that, with fresh water making up only about 2.5%. Of the fresh water, only 0.3% is in liquid form on the surface.

Obviously, there is a disproportionate amount of salt water and a shortage of fresh water on the earth. It would be nice if we could take the salt out of some of that salt water so it could be used. And, in fact, you can. Using a still.

Probably the leading technology experts on removing salt from the water today is the small country of Israel. They have become a major water exporter with innovative techniques for conservation and desalination. You would probably enjoy reading "Let There Be Water: Israel's Solution for a Water-Starved World" by Seth M. Siegel.

However, you can desalinate water yourself with a solar still. You will need two containers, one smaller in diameter than the other, and lower in height. Place the small container within the larger container. Add salt water to the larger container, but leave the small container empty. Place a plastic sheet of some kind plastic wrap tightly over the larger container. Fasten it tightly. Add a small rock, marbles, or beads to the center of the plastic wrap so that it creates a concavity inward. Set this apparatus in the sun for a few hours and water will evaporate, condense on the plastic wrap and drip into the smaller vessel. This water that accumulates there is fresh water with no salt.

> You cannot drink the ocean
> And it's all Gods fault
> He could have put in sugar
> But he went and put in salt

THE SOLAR SOLUTION

Chapter 10 - NEW FANGLED ELECRICALIFICATION

There is a house in New Orleans
They call the Rising Sun
And it's been the ruin of many a poor boy
And God, I know I'm one
My mother was a tailor
She sewed my new blue jeans
My father was a gamblin' man
Down in New Orleans
Now the only thing a gambler needs
Is a suitcase and trunk
And the only time he's satisfied
Is when he's on a drunk
Oh mother, tell your children
Not to do what I have done
Spend your lives in sin and misery
In the House of the Rising Sun
   - Traditional -

Part 10.1 - Solar-Powered, Thermoelectric Generator

Dear Earth,
   I hope you are enjoying your stupid DAY.
Sincerely, Pluto
   You can understand that attitude from a planet that was only recently demoted. The other planets probably feel left out too. It does seem a little cheeky to have an "Earth Day" when there are many other perfectly fine planets with no "designated day" to celebrate them. Maybe a planet has to be despoiled before it warrants a day. Maybe it's like honoring dead men at their funerals whom we treated badly when they were alive.
   The best thing I have done for the earth, so far, is to retire. Boy, that sure lessened my carbon footprint. Maybe when everyone is unemployed, we will save the earth! I do feel badly because I sold my old truck, instead of junking it, when I had the chance back with the cash-for-Junkers stimulus. That means someone is still driving that old clunker around. The trees liked it, though. I remember them inhaling when I would drive by.
   But I digress. I do that a lot. . . My grandson and I just finished

a prototype for our solar-powered, thermoelectric generator. I think I will name it SPTG to avoid having to write that out every time. I guess I could name it Charlie, but SPTG is shorter. Anyway, it's not that SPTG's hadn't already been invented. But ours is a unique design: and well, it's ours. . .

We attached to twelve radiating aluminum wires to a graphite and zinc center using a zinc washer about the size of a fifty-cent piece. I suppose the washer has a gauge, but we just scrounged it out of my "things" drawer. We needed some graphite, so I took a carpenter's pencil and sanded the wood off the lead before wiring it to the zinc washer.

We used aluminum wire for the rays. These wires attached to the washer and graphite in the center, then looped away from the center and back again like flower petals. When they looped towards the center, they bent and connected to the next flower petal so that there was a continuous circuit from the center to each petal. Then they came back in and are attached to the next petal. The last two aluminum petals were not connected to each other unless we connected them with a voltmeter.

It worked! When we heated the graphite center with an alcohol lamp and connected the last two aluminum petals with a voltmeter, we got a voltmeter reading of about 16 millivolts. No, really! Sixteen millivolts! Just 986 more, and we'll have a whole volt. No, not a car. An electrical charge that's almost as large as a AA battery. Pretty impressive, huh?

What happens is that, as the graphite center heats up, the electrons move around. The heat pushes some of the electrons out towards the tips of the aluminum petals where the aluminum wires cool. Anytime you have a flow of electrons; you have an electrical current. Since all the petals are being heated at the center and are connected to one another, the electrons flow along the wires until the circuit is interrupted by not having two of the petals connected.

When we attached a voltmeter to the two unconnected petals, we complete the circuit and could measure the current flow. The current flow is relative to the heat at the center and to how cool the petals are. The greater the heat differential, the greater the current.

Therefore, the more petals there are, and the larger they are, the greater the cooling capacity. Now when we placed the device into a box where we focused the sun on the central zinc and graphite core, we got a much greater source of heat. By shading the petals from the sun, we increased the differential and created almost 112 mv. The illustration below shows the crude sunflower as powered by an alcohol lamp. With an entire array of these "sunflowers," we should be able to save the planet from something. Then we can turn our attention to the problem of planet-envy and see what we can do about getting Pluto its own day.

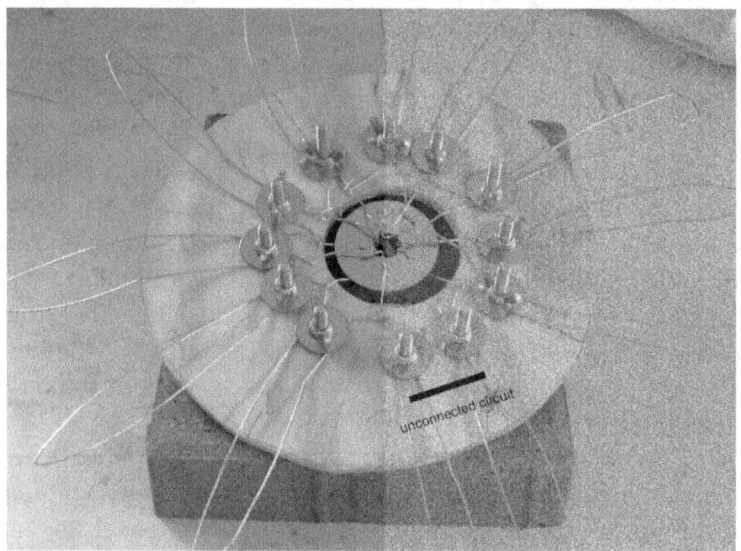

Part 10.2 - Photovoltaic Nightlights

> The foolish gardener planted a light bulb and thought he would get a power plant.

    Storing up solar energy to use at night gives a whole new meaning to "daylight savings." The daylight savings they talk about every winter and spring doesn't save any daylight at all. At least it doesn't save any light you can retrieve for short winter days when one could use a little more light.
    No. This idea is much better. There are some solar, pathway lights that use photovoltaic collectors to store light during the day and convert it back into light at night. These are often used to light sidewalks to homes. Really? Who cares about sidewalks? They just make it easier for burglars to find the windows!

I am much more concerned about the monsters under my bed, or the middle of the night trips to the restroom. Now there's a problem someone could shed a little light on. But why pay the electric company for your convenience from nightmares and stubbed toes in the dark? Here is what you can do to save some daylight.

Like a lot of science, we begin by taking something apart. So, buy one of those inexpensive solar pathway lights. The solar collector and light source are set on top of a stake that is normally planted into the ground. Like a lot of science, we start by doing things abnormally, so take the top off the stake. Usually, you can do this by just twisting it off. Then remove the little strip of paper that protects the battery and place the collector in the sun for the day.

In the evening, place the solar panel into the mouth of a jar of some kind. A wide-mouth canning jar of any size should work, but you can also use any glass container. If you choose the jar to fit the solar collector, it will be easy to fit the collector snuggly into the jar. If it doesn't fit, you may have to figure out some way to attach the collector so that light doesn't escape out the top. The idea is to diffuse the light through the dark, so it gives a muted glow.

Voila! Your own solar-powered, daylight-savings device.

Of course, you could glue opaque figures on the sides of the jar to cast shadows, use colored jars to create mood, etch the glass to mute the light further, or in numerous other ways express your artistic talent in your science project.

> When the lights went out, Leroy re-fused to put the power back on.

Part 10.3 - Solar Windmill

The Ingenious Gentleman Don Quixote of La Mancha or, as the say in Spain, El ingenioso hidalgo Don Quixote de la Mancha, is a Spanish novel by Miguel de Cervantes Saavedra. Probably the most enduring image of the character representing his quest to reinstate chivalry, is of him tilting at a windmill.

You might find recent arguments and discussions regarding wind energy similar. The problem is it isn't clear which side is tilting at the windmill. Is it those who are trying to make wind energy an efficient industry, or those who oppose windmills?

Of course, wind energy is actually solar energy. Winds are created by warmed and cooled air masses. When the two meet, the cold air sinks because it is denser. It slides under the hot air, creating a wind. The hot air rises because it is less dense, and this create updrafts, as well.

So windmills must be constructed where topography and prevailing weather create areas of sustainable, or at least frequent, winds. Unfortunately, areas appropriate for windmills are not found everywhere, so land on which to build windmills is a limiting factor.

But what if we could generate a source of hot air that would give us sustainable winds wherever there was enough sunlight? Well, it can be done. We can build a solar chimney. A trial model can be built by opening both ends of several aluminum cans and taping them together to make a metal tube.

Set this tube on a stand where air can enter through the bottom of the tube. Placing the column between two book s or blocks of wood would work fine.

Use a malleable wire to erect an arch across the top of the chimney. A large, unbent large paper clip works well for this, as would any other relatively soft wire. Attach a thumbtack or straight pin to the center of the wire with the point directed upward.

Now, it's finally time to create a windmill at which to tilt. Obtain a six-inch square piece of paper. Cut in diagonally towards the center from each corner to within a quarter of an inch of the center of the paper. Bend every other point from the corners to the center and tape them in place. Then place the pinwheel on top of the thumbtack, so that it is balanced and centered.

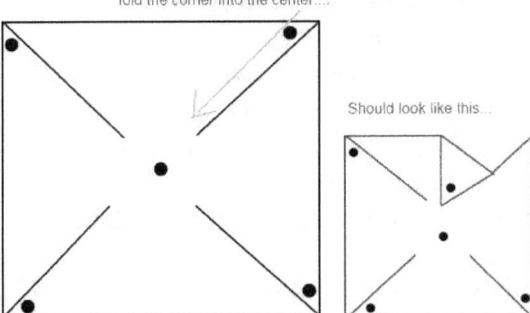

Now place the tower in the sunlight. The heat from the sun should warm the air inside the cans to create an updraft causing the wheel to spin.

There are a number of experiments to perform with this apparatus.

1. If the wheel doesn't spin too quickly, you might count the number of revolutions per minute. Then paint the cans black on the outside and redo the experiment to see if the wheel turns faster.

2. You could also measure the heat inside the chimney with a thermometer. See what difference various colors of paint might make to the inside temperature.

3. Set the can flat on a surface, or in other ways reduce the gap at the base of the chimney, to see if that changes the air flow out the top.

4. Can enough up draft be created to create an electrical current?

> I call a power failure a current event.

Chapter 11 - SOLAR REFRIGERATORS

All day I've faced the barren waste
Without the taste of water, cool water
Old Dan and I with throats burned dry
And souls that cry for water, cool, clear water

Keep a-movin', Dan, don't you listen to him, Dan
He's a devil not a man
And he spreads the burning sand with water
Dan can you see that big green tree
Where the water's runnin' free
And it's waiting there for you and me

The nights are cool and I'm a fool
Each star's a pool of water, cool water
But with the dawn I'll wake and yawn
And carry on to water, cool, clear, water

The shadows sway and seem to say
Tonight we pray for water, cool, water
And way up there He'll hear our prayer
And show us where there's water, cool, clear, water

Dan's feet are sore he's yearning for
Just one thing more than water, cool, water
Like me I guess he'd like to rest
Where there's no quest for water, cool, clear, water
    Bob Nolan

    You know what would be cool? If you could use solar energy to keep food cool. Places that don't have electricity for stoves usually don't have refrigerators either. Consequently, a lot of illness can occur because of improper storage of food, and food must be cooked in only small amounts so there are no left-overs.
    You could use one of more of the methods in the previous chapter to make electricity to run a cooling unit. But that may not be possible

for economic reasons. But solar energy can sometimes be used to at least cool some food to make it more refreshing or to keep fresh vegetables and fruit from spoiling too quickly.

Because of waters high specific heat, it is difficult to heat to boiling without supplying a lot of energy. But when water cools, it also must use a lot of heat to evaporate. If one can evaporate water form a close system, it will cool the inside of the container significantly.

This only works if the external and ambient humidity is low. If the humidity is high in the surrounding, the water in the container will not evaporate efficiently, or at all.

There are several ways to make a solar cooler. The simplest would be to erect a frame over which to drape cloth of some kind. If the cloth is soaked with water and set in the sun, the water will slowly evaporate cooling the inside of the container. Of course, the cloth would have to be kept wet, which could be tedious in hot dry conditions. Further in such conditions the water required might be excessive and it might be better used to sustain life.

A more substantial type of solar cooler can be built that takes less oversight and less water by nesting two containers, one inside the other. It is best if the two containers are at least partly porous. I have used two large clay flower pots, or as pictured here, cinder blocks.

Place them so there is a central space, then fill the gaps between the inner and outer vessels with fine sand. Fill the space with the sand with water. Cover the central opening with an insulated cover of some kind, that still allows water to evaporate from the sand.

As the sun strikes the outer container it heats it and the water in the sand begins to evaporate, thereby cooling the inner container and the space in the center. This creates a small cooler effect inside and will help keep raw fruit from spoiling as quickly, or cool liquids to make them more refreshing to drink.

Chapter 12 - TIME FOR SOLAR COOKING

Sally go round the sun
Sally go round the moon
Sally go round the chimney top
Every afternoon
   - Folk Song -

Part 11.1 - Time of Year

A thousand years ago the Anasazi Indians of the American Southwest used the sun to determine the seasons. Lest you misunderstand, I was not alive then as some people have rumored. I am taking the word of the people who administer the Chaco Canyon National Historic Park, your National Park Service. They tell me Chaco Canyon in New Mexico was inhabited by these Indians until about 700 years ago.

When the Anasazi left the area, they left behind an amazing solar clock called the "Sun Dagger." On the side of Fajada Butte, a monolith was raised about 400 feet high from the desert floor. There were positioned two slabs of rock placed near the top of the butte. These rocks cast a shadow against the cliff wall, and a spiral petroglyph was carved into the sandstone. This carving is in shadow most of the time.

However, on the summer solstice, a sliver of light shows between the slabs, first at the top of the spirals and then descended through the spirals, slicing them in half. This happens over a period of about 18 minutes.

During the winter solstice, two daggers of light appear on either side of the spiral exactly framing it.

The spiral carved in the rock has nine grooves. During spring and autumn equinoxes, the shaft of light called a dagger, cuts through the spiral exactly between the fourth and fifth grooves. It's spooky. It's almost like it was on purpose or something!

There are a large number of these light displays tracking the seasons throughout the United States. They can be found at Hovenweep National Monument, Burro Flats in California, and La Rumorosa in Baja, California. The Bighorn Medicine Wheel in Wyoming is also a solar calendar of sorts. Near present-day St. Louis, there existed a Mississippian culture that also had Medicine Wheels. One of their Medicine Wheels has been rebuilt and can be seen at Cahokia Mounds State Historic Site.

Obviously, there are certain times of the year that are better for cooking with solar than others. Is it just a coincidence that those seasons are determined by the sun? Of course, we divide the year into months based on the phases of the moon. But the phases of the moon only exist in relation to the sun.

> *"The first grand discovery was time, the landscape of experience."*
>
> *- Daniel Boorstin -*

On the University grounds where I taught for many years, there was a metal model of a rock monolith in the area. Metal bars were placed on the ground next to it such that, at the summer solstice in our region the shadow of the metal bars exactly matched the bars on the ground. Most students ignored it and didn't even know what it was. With central air and heating, the sun becomes a bit of a distraction known mostly for causing skin cancer.

Part 11.2 - Time of Day
> *Does anyone really know what time it is?*
> *Does anyone really care?*
>   Chicago

THE SOLAR SOLUTION

One has to know what time it is if cooking by the sun. One should probably know what time it is if you're planning to eat in the day time, or planning when to go to bed. "Early to bed and early to rise makes a man healthy, wealthy, and wise." "Poor Richard" said that, so I am not too sure he knew much about being wealthy.

Of course, "Poor Richard" was Benjamin Franklin, and he was both wealthy and wise. In fact, he's the guy who invented daylight savings time! He published a paper entitled "An Economical Project for Diminishing the Cost of Light" in The Journal of Paris in 1784. In the essay, he suggested, that Parisians economize candle usage by getting people out of bed earlier in the morning and by making use of the natural morning light instead.   Surely, suggesting that Parisians go to bed early, and get up early, had to be a joke.

Well, look who's laughing now. Not me! I hate daylight savings time. But I don't mind using the sun to tell the time of day. It's actually kind of fun. Although it's not as simple as checking your cell phone. All you need is a sundial.

Sundials are easy to construct, although the easily-constructed ones usually don't tell the correct time. It's trickier than it sounds, as seems to be the case with just about everything. The reason it's a little tricky is that while the sun does appear to revolve around the earth, the earth is actually revolving around the sun.

The earth is a sphere, although not a perfect one. It is kind of oblong. This makes the sun appear as if it is in different parts of the sky as seen from the earth. So the position of the shadow not only shifts with the time of day, but with the season of the year. Not only that but the latitude at which you live also dictates the placement of the sundial. A sundial set at 20° latitude will not be accurate at 60° latitude.

You guessed it. Most of those beautiful sundials advertised for sale in garden magazines won't give you accurate time where you live, unless you accidentally live at the same latitude in which it was made. That place is probably somewhere in China. About half of China is either above or below the latitudes encompassing the United States, so the odds aren't good.

There are a lot of places on the internet, and books in your local library, that will tell you how to make a simple sundial from a paper plate or other round object. But what if you really want to know what time it is, not just based on some dumb, old, oscillating, atomic nucleus or something? Well, you can do that, but you have to really want to.

Actually, "sundial" is not its real name. It is a "summer sundial" because it only works from the spring equinox until the fall equinox. During the winter months, at most latitudes, the sun is too far north or south to project a shadow.

The sundial works because the pole that is used to cast a shadow is parallel to the earth's axis. The face of the sundial should be parallel to the equator. If you are at the North Pole, the vertical pole would be a straight up and down pole and the hours would be marked evenly around the base. However, the odds of that are small, so you will have to do a little calculating and arranging.

Gnomons

By the way, the pole of the sundial is called a gnomon. I know that sounds like a troll living under a bridge or something. It comes from Greek and anciently meant a discerner or an interpreter. That seems vaguely appropriate.

The gnomon can be made of many different materials: wood, metal, PVC pipe, even antlers. It probably shouldn't be much more than three inches long. You need to be able to mount this in the center of your sundial face, but your sundial face will also be bolted down to the post. So making some mechanism for fitting over the bolt head and then making the gnomon so that it slips over that would be ideal. For example, you can mount the gnomon on another smaller piece of wood and attach it over the bolt hole.

If you were using a 5/16-inch by a 2-inch lag screw to attach the face, drilling a piece of wood that slips over the head would work fine. This screw could be the gnomon, or you could then drill some other object out to slip over the piece of wood. The gnomon is where the artistry comes in, so I am being purposely vague. I don't want to inhibit your creativity, and I don't have any to spare.

Mounting

Making the post on which the sundial sets is at least as important as making the face of the sundial and gnomon. That explains why fancy, commercial sundials are seldom correct. They generally come with no instructions about proper mounting,

A 4 by 4 post, pressure formed and outdoor-treated wood is probably the best. It should also be straight, without cracks that might split when a hole is eventually drilled in it.

The post length is not too important except that you don't want it sticking up four or five feet in the air. At that height, you would be looking into the sun at times when trying to read it. Shorter is better, but it does need to be set solidly in the ground, by a foot or two, in cement.

The top must be cut at an angle to mount the face correctly. The angle will vary with latitude. You can find the correct angle by subtracting your latitude from 90°. In Grand Junction, Colorado you would subtract 39.0639° N from 90 degrees. If you are not obsessive-compulsive, you could just subtract 40° which would make the needed angle 50°.

If you make a line at right angles to the post about four inches from the top using a carpenter's square, you can then use a protractor to mark your 50-degree angle. Make the right angle line the bottom of the angle. This is a critical cut, so measure twice and cut carefully. Now locate the center of the sloping face and drill a hole a couple of inches deep at that point.

You can place the post wherever you want, but sundials work best if they are in full sun. Dig your hole deep enough to anchor the post securely because, if it shifts, you will have a hard time telling time. This is why it is a good idea to set the post in concrete.

The slanted angle must face true north. You can use the north star for a guide if you don't mind working late, but a good compass probably works best. The post must be set perfectly vertical using a plumb bob or carpenter's level.

Face

The face of the sundial can be any size you want. However, it is a little easier to read if it is large. It should be made of materials that can stand up to weather, so something like 3/4-inch plywood works well. A twenty-inch diameter would be a nice size. The face can be square, although traditionally, the sundial face is round.

You'll probably want to paint it with a primer as well as weather-proof paint to protect it from the elements. Traditionally, sundials have ornate decoration and numbers, although that isn't necessary. If you have an artistic bent you could certainly decorate the face, however, you wish.

Placing the numbers is critical. Place the number twelve at whatever point you want to call the top. Find the center of your face and draw a line from the twelve o'clock to the center point. Use a protractor to move exactly 15 degrees to the right, and mark the number one there using a straight edge to draw a line to the center. Continue moving and marking each number exactly 15 degrees to the right in a clockwise direction.

The next number twelve must be exactly opposite of the first twelve on your board or where the six o'clock would be on a clock face. Continue marking hours at every 15 degrees to the right until you have completed the clock face.

Mounting the Sundial Face

Drill a hole in the exact center of the sundial that will accommodate your mounting screw. You will probably want to use a washer to protect the center hole. Mount the face and tighten the screw only tight enough to hold it in place, but still loose enough to be able to turn the face.

Now set the time. Rotate the sundial face with your right hand until the shadow of the gnomon on the sundial face reads the same as the clock time. This must be done during Daylight Saving Time. Mark the position of several points to drill holes to keep the face from rotating. Tighten the lag screw firmly without moving the sundial face. Drill holes for the screws that secure the face without letting it change position.

Part 12 - EPILOG

It's time to draw this discussion of personal solar energy to a close. Remember,

> "He makes his sun rise on the evil and on the good,
> and sends rain on the just and on the unjust."
> - Jesus Christ -

> "From the rising of the sun to its setting the name
> of the LORD is to be praised."
> - Psalm 113:3 -

> "But for you who fear My name, the sun of righteousness
> will rise with healing in its wings; and you will go forth

and skip about like calves from the stall."
- Malachi 4:2

"Then the righteous will shine forth like the sun in the kingdom of their Father. He who has ears, let him hear.
- Mathew 13:43

I believe in Christianity as I believe that the sun has risen: not only because I see it, but because by it I see everything else.
- C. S. Lewis -

APPENDICES

Appendix A – Plans for solar cookers

Box Cooker

Boxes: Box cookers can be manufactured in many ways and from many materials. Very functional and easy-to-make cookers can be manufactured with cardboard, tape, and glue. As scientific as all this sounds, it ain't rocket science. Many manufactured, commercial boxes are made of various kinds of plastic. They can also be made of wood or metal if a person has the desire and skills.

The cooker consists of two boxes, one inside the other. The dimensions of the outer box are not very significant. The outside container must be at least a couple of centimeters, half an inch, or more, larger than the inner container. You are going to fill the gap with insulation, newspaper, or other inexpensive inert material. Dead air space works almost as well.

The dimensions of the inner box need not be rigid, although the inside dimension should probably be at least 40 cm x 40 cm, approximately 16 x 16 inches. Size will partly be determined by the size of your cookware. A larger box can be more flexible in the amount you are able to cook; but if you are only cooking small amounts, a larger box takes longer to heat up.

Your oven's inner box should be about 1 inch, 2.5 cm, deeper than your largest pot and about 1" shorter than the outer box so that there will be a space between the bottoms of the boxes once the cooker is assembled.

Lids: The lid of the two boxes should be slightly larger than the outer box so that it slips over the outside edge and hangs down over the boxes by at least two or three inches. You will cut out the center of the lid and replace it with a transparent material. Cellophane isn't the best because it tears easily, sags when hot, and melts if it touches the hot cooking pan. Turkey-baking bags work well, but they must be replaced periodically because of damage by UV light. Various plastics also work with varying degrees of success. Glass works extremely well, of course, but is expensive, heavy, and breakable.

Reflectors: Box cookers often need reflectors to be most effective. These can be constructed from cardboard, and a great reflective surface can be made of tin or aluminum foil.

Absorbing Material: The inside of the box needs to be painted black. Be sure to use a non-toxic, flat paint or tempura. Actually, fire ashes mixed with flour and water make an adequate absorbent color, if a little messy.

Funnel Cooker

Funnel: Funnel cookers are made like funnels. Any material can be used, but the skill level for making a funnel is a little more difficult. I have found cardboard, covered with tinfoil, works well. Mylar works well if you can make a rigid frame for it. A piece of material about 2 feet by 4 feet long will give you a funnel size that will cook one pot at a time. The funnel can be held upright, quite simply, by placing it in an appropriately sized cardboard box or even in a hole in the ground.

You will need a platform at the bottom of the funnel on which to set the cooking vessel. I use a stiff, wire screen, cut in a circle, that sits above the very bottom of the funnel.

Lids: It is difficult to put a lid on a funnel. However, what works well is to surround the cooking vessel with a roasting bag like those used for roasting turkeys and hams. This plastic retains heat without burning and can be reused several times with care.

Reflectors: The inside of the funnel is lined with tin or aluminum foil as a reflecting surface. This works well because it wrinkles and does not present a uniform surface. This focuses light onto the cooking vessel without creating overheated areas.

Absorbing Material: Using a funnel cooker, you will want to use blackened cookware. Heavy steel pots absorb heat but require an awful lot of heat to attain cooking temperatures. They can also be too heavy to be supported by the cardboard funnel. A glass canning jar, painted externally black, works well. You can also purchase inexpensive pots and pans with lids that are dark in color.

Pizza Box Cooker

Box: A recycled pizza box can be made into a small solar cooker.

Lid: Draw a one-inch border on all four sides of the top of the pizza box. Cut along three of these lines, leaving the line on the back of the pizza box uncut. Form a flap by gently folding back along the uncut line to form a crease. Tape a piece of plastic, cellophane, or other clear material over the opening created when you cut the flap in the pizza box. Be sure the plastic becomes a tightly sealed window, so that the air cannot escape from the oven interior.

Reflector: Cut a piece of aluminum foil the size of the flap. Smooth out any wrinkles and glue it into place on the part of the flap that would face down if you shut it.

Absorbing Material: Cut another piece of aluminum foil to line the bottom of the pizza box, and carefully glue it into place. Cover the aluminum foil with a piece of black construction paper and tape this into place covering the foil. The food can be placed directly on the black construction paper, on a piece of foil, a wire screen, or any other slightly raised surface.

Cooking: Close the pizza, box-top window, and prop open the flap of the box with a wooden dowel, straw, or other device. Then face the box towards the sun. Adjust it until the aluminum reflects the maximum sunlight through the window into the oven interior. A pizza box cooker is great for cooking flatbreads, cookies, rolls, thin meats, and vegetables.

Appendix B – Cooking Guidelines

General:
- Most recipes take slightly less liquid when cooked in a solar oven.
- Cooking time depends on the temperature of the food as it is placed in the oven, as well as the brightness of the day.
- Allow plenty of time for cooking. Foods hold well in the solar oven without scorching or drying out.
- Check food about once an hour when you're a beginner getting started.
- Most recipes calling for at higher temperature will do fine if given more time to cook.
- Set the food out early and don't worry about overcooking.

Types of solar cooking days:
- GOOD: Clear and sunny. Oven will preheat to 275° - 300°F (130° - 145° Celsius.)
- FAIR: Hazy or partly cloudy. Oven will preheat to 200° - 225°F (95° - 105° Celsius.)
- BAD: On a completely cloudy day one cannot cook with solar heat.

General Cooking Times:
- **Vegetables** (Potatoes, carrots, squash, beets, asparagus, etc.)
    - *Preparation*: There's no need to add water when cooking fresh produce. Cut items into slices or "logs," to ensure uniform cooking. Corn cooks fine on or off the cob.
    - *Cooking Time*: About 1.5 hours
- **Cereals and Grains**: Rice, wheat, barley, oats, millet, etc.
    - *Preparation*: Mix 2 parts water to every 1 part grain. The amount may vary according to individual taste. Let the grain soak for a few hours for faster cooking. To ensure uniform cooking, shake the jar half way through the cooking time.
    - *Cooking Time*: 1.5-2 hours
- **Pasta and Dehydrated Beans and Soups**
    - *Preparation*: First heat the water to near boiling. This should take between 50 and 70 minutes. Then add the pasta or soup mix and stir or shake, and cook 15 additional minutes.
    - *Cooking Time*: 65-85 minutes

- **Beans**
  - *Preparation:* Allow tough or dry beans to soak overnight. Place beans in a cooking jar with water to cover.
  - *Cooking Time:* 2-3 hours.
- **Eggs**
  - **Preparation:** There is no need to add water.
  - *Cooking Times:* 1-1.5 hours, depending on yolk firmness
- **Baking**
  - *Preparation:* Times vary based on the amount of dough being baked.
  - *Cooking Times:* Bread: 1-1.5 hours; Biscuits: 1-1.5 hours; Cookies: 1 hour
- **Meats:** Chicken, beef, and fish
  - *Preparation:* There is no need to add water. Longer cooking times makes the meat more tender.
  - *Cooking Time:* Chicken: 1.5 hours cut up or 2.5 hours whole; Beef: 1.5 hours cut up or 2.5-3 hours for larger cuts; Fish: 1-1.5 hours
- **Quick-Cooking Foods** started early will be done early on a fair day: - Rice, whole grains, rolled grain flakes, cereals, most egg dishes, chops, ribs, fish and most poultry, puddings, crackers, cookies, brownies, fruits, green vegetables, shredded vegetables.
- **Medium Cooking Foods** if started early, ready by noon on a good day or by evening on a fair day: cornbread, gingerbread, medium-sized roasts, quick bread, yeast rolls and buns, soufflés, root vegetables such as potatoes, turnips, some beans such as lentils, black-eyed peas, black beans.
  - **Longer Cooking Foods** if started early, ready for late lunch or dinner on a good day: whole turkey, large roasts, stews and soup and bean pots (unless brought to a boil before placing in solar oven), most yeast breads and cakes, pre-soaked pinto beans, field peas, garbanzo beans, small navy beans, soybeans, kidney beans, red beans, yellow peas, dried peas, split green or brown peas.

Appendix C – Selected Recipes

The following recipes are courtesy of The Solar Cooking Archive and are available at: http://solarcooking.org

Most of these recipes were developed using simple solar box cookers which cook at temperatures between 250F and 300F.

<u>Solar Oven Lasagna</u>
1 32-oz. jar spaghetti sauce
1 pound ricotta cheese
1 pound mozzarella cheese, shredded
Parmesan cheese
8 oz. package of lasagna noodles

Spread 1 1/2 cups of sauce over bottom of dark roaster. Coat uncooked noodles with ricotta cheese and layer over the sauce. Add half of the mozzarella cheese. Repeat layers of sauce, noodles and cheese. Top with remaining sauce. Sprinkle Parmesan cheese over the top. Cover and bake for 3 hours. 1 pound of ground beef cooked in a separate dark pan may be added to the sauce before preparing the lasagna.

<u>Easy French Bread</u>
1 package yeast
2 cups water
4 1/2 cups white flour
1 tablespoon sugar
2 teaspoons salt

Dissolve yeast in one cup lukewarm water. Sift flour with sugar and salt into a large bowl. Stir in dissolved yeast. Add just enough of the second cup of water to hold dough together. Mix until dough is sticky. Cover with a cloth and let rise until doubled. Butter or grease a round roaster and add dough to dark pan. Let rise another half hour. Cover. Bake in solar oven until golden brown, about 2 hours.

Chicken in the Pot
4 chicken breasts, halved and skinned
4 medium potatoes, quartered
2 large carrots, cut into 1-inch chunks
2 stalks celery, cut diagonally into 1-inch chunks
1 can Swanson's chicken broth
1/4 teaspoon pepper
pinch of basil
pinch of rosemary

Place chicken in a 3-quart pot or casserole. Arrange vegetables over the top. Sprinkle with seasonings. Add chicken broth. Cover and cook approximately 1 1/2 to 2 hours. Stir a couple of times while baking. You may substitute thighs for chicken breasts and add other seasonings.

Split Pea and Potato Soup
1 cup split peas (frequently sold in bulk bins)
1 bouillon cube (chicken, beef, vegetable, etc.) or 1 teaspoon bouillon broth powder
1/2 cup diced potato pieces

Put peas, potatoes, and bouillon in 32 oz. black-painted mason jar. Or, if using a 26 oz. jar, use only 3/4 cup of split peas. Fill the jar with water to within 1 inch of the neckline. Seal with the black-painted ring and lid (oil the inside parts of the lid and ring first). Cooks in 2.5 to 8 hours depending on conditions, solar cooker type, etc. Note that texture of soup will vary with cooking time. Thoroughly cooked pea soup will have almost no pea chunks in it -- they all dissolve! So, watch the soup carefully after the first 1.5 hours if you like your soup with a few remaining split peas.

Solar Sweet and Sour Chicken
1 15-oz. can sweet and sour sauce
1 8-oz. can pineapple chunks or tidbits, drained
1 chicken breast or 2 chicken legs

If using chicken breasts, cut them in half and remove skin. Cut into small chunks. Place in a dark pan. Add sauce and pineapple. Cover and bake for 2 to 3 hours. Serve over brown or white rice.

Sunshine Chili
1 pound small red beans (dry)
1 pound ground chuck
2 medium onions
1 small green pepper
1/2 cup fresh parsley, minced
dash of salt and pepper
1 28-oz. can tomatoes (may be blended first)
4 cups V-8 juice
1 tablespoon chili powder

Soak red beans in water overnight. Brown ground chuck, onions, green pepper, parsley, salt, and pepper. Drain well after about one hour. Add tomatoes, red beans, V-8 juice and chili powder. Cover and bake about 4-5 hours in the solar oven; serves 10. You may add more V-8 juice as cooking proceeds.

Solar Brownies
1/2 cup shortening
2 1-oz. squares unsweetened chocolate
2 eggs
1 cup sugar
1 teaspoon vanilla
3/4 cup flour
1/2 teaspoon baking powder
1/2 teaspoon salt
1 cup broken walnuts

Melt shortening and chocolate together in solar cooker; cool. Beat eggs until light; stir in sugar, then chocolate mixture and vanilla. Add dry ingredients, mix well. Add nuts. Bake in greased 9-inch round dark roaster pan, covered, for one hour. Cut into squares.

Solar Crustless Apple Pie
6 apples (chopped or sliced)

1/3 cup sugar
Cinnamon
1/2 cup butter
1 cup flour
1 cup brown sugar

Place apples and sugar in buttered round or oval black roaster. Work together with the flour, brown sugar, and butter and sprinkle over apples. Sprinkle cinnamon on top. Cover and bake in solar oven about two hours. Cool, serve with vanilla ice cream (non-solar product).

Lunar Chiller Recipes
These are not sanctioned by any other solar related agencies, or anyone with good sense.
Breakfast Iced Coffee
I don't have a clue how to do this. ;-)

Breakfast Iced Tea
Put tea or tea bags in glass or clear plastic canister. Set this in the sun to steep all day. After dark place in lunar Chiller for the night. Iced tea in the morning.

Chilled Breakfast Milk
Milk the goat, cow, horse, alpaca, whatever in the evening just before sundown. Immediately place the milk in the lunar chiller at dark.

THE SOLAR SOLUTION

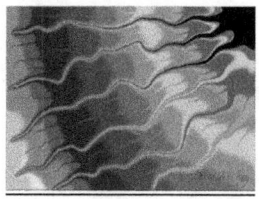

Grab your coat and get your hat
Leave your worries on the doorstep
Life can be so sweet
On the sunny side of the street

Can't you hear the pitter-pat
And that happy tune is your step
Life can be complete
On the sunny side of the street

I used to walk in the shade with my blues on parade
But I'm not afraid...this rover crossed over

If I never had a cent
I'd be rich as Rockefeller
Gold dust at my feet
On the sunny side of the street

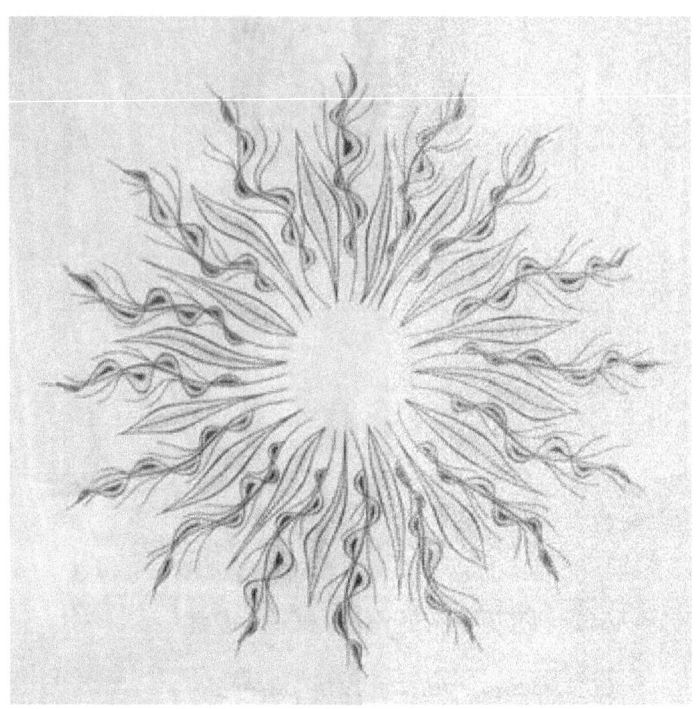

STORMY WEATHER

*Don't know why there's no sun up in the sky, stormy weather
Since my gal and I ain't together, keeps raining all the time
Life is bare, gloom and misery everywhere, stormy weather
Just can't get my poor old self together
I'm weary all the time, the time
So weary all the time*

*When she went away, the blues walked in, and they met me
If she stays away, that old rocking chair's gonna get me
All I do is pray the Lord above will let me
Walk in the sun once more*

*Can't go on, everything I have is gone, stormy weather
Since my gal and I ain't together Keeps raining all the time
Keeps raining all the time*

OTHER BOOKS BY GARY MCCALLISTER

MUSIC
Making More than Music 2014

First Songs with the Mountain Dulcimer: history, instrument, and simple songs 2015

Hymns on Mountain Dulcimer: Learn to play the mountain dulcimer using hymns 2016

SCIENCE
Hanging Out With GRAVITY: Galileo's gravity game 2015

Seriously Silly Science: A Science reader for the whole year – and some of it is even true 2015

A Convenient Truce: A cease-fire in the war between religion and science 2016

NOVELS
Walking Man 2015

The Solar Solution

Gary McCallister is Professor Emeritus at Colorado Mesa University where he was an award-winning teacher, author, and digital developer for many years. He was awarded the Distinguished Faculty Award in 1988. In 2012, he received a state-wide award for his science column in the local paper, and in 2004 he received the Communicator Award for an instructional CD on teaching methods. His interest in solar cooking and water sterilization stemmed from his research involving parasitic diseases.

www.ingramcontent.com/pod-product-compliance
Lightning Source LLC
Chambersburg PA
CBHW060408190526
45169CB00002B/811